WARRIOR • 156

EARLY ROMAN WARRIOR 753–321 BC

NIC FIELDS

ILLUSTRATED BY SEÁN Ó'BRÓGÁIN

Series editor Marcus Cowper

First published in Great Britain in 2011 by Osprey Publishing
Midland House, West Way, Botley, Oxford OX2 0PH, UK
44-02 23rd St, Suite 219, Long Island City, NY 11101, USA
E-mail: info@ospreypublishing.com

OSPREY PUBLISHING IS PART OF THE OSPREY GROUP

© 2011 Osprey Publishing Ltd.

All rights reserved. Apart from any fair dealing for the purpose of private study, research, criticism or review, as permitted under the Copyright, Designs and Patents Act, 1988, no part of this publication may be reproduced, stored in a retrieval system, or transmitted in any form or by any means, electronic, electrical, chemical, mechanical, optical, photocopying, recording or otherwise, without the prior written permission of the copyright owner. Inquiries should be addressed to the Publishers.

Every attempt has been made by the Publisher to secure the appropriate permissions for material reproduced in this book. If there has been any oversight we will be happy to rectify the situation and written submission should be made to the Publishers.

A CIP catalogue record for this book is available from the British Library

ISBN: 978 1 84908 499 4

E-book ISBN: 978 1 84908 500 7

Editorial by Ilios Publishing Ltd, Oxford, UK (www.iliospublishing.com)
Page layout by: Mark Holt
Index by Marie-Pierre Evans
Typeset in Sabon and Myriad Pro
Originated by PDQ Media
Printed in China through Worldprint Ltd

11 12 13 14 15 10 9 8 7 6 5 4 3 2 1

Osprey Publishing are supporting the Woodland Trust, the UK's leading woodland conservation charity, by funding the dedication of trees.

www.ospreypublishing.com

ARTIST'S NOTE

Readers may care to note that the original paintings from which the colour plates in this book were prepared are available for private sale. All reproduction copyright whatsoever is retained by the Publishers. All enquiries should be addressed to:

Seán Ó'Brógáin
Srath an Ghallaigh
An Clochan
Leifear
Tir Chonaill
Ireland

The Publishers regret that they can enter into no correspondence upon this matter.

CONTENTS

EARLY ROMAN WARRIOR *c.*753–*c.*321 BC

Ludovisi Mars Ultor, Mars the Avenger (Rome, MNR Palazzo Altemps, inv. 8654). Rhea Silvia, the mother of Romulus and Remus, was of the bloodline of Aeneas, the ultimate ancestor of the Romans. She claimed that the conception of her twins was a divine event, the father being Mars. Whatever the reality of their origins, this fantastic story, which intermingles the human and heavenly, expresses how the Romans chose to view and represent themselves, namely as the embodiment of their inborn strength and their overwhelming superiority in warfare. (Fields-Carré Collection)

INTRODUCTION

We are unlikely to ever know for certain whether Romulus really lived, but regarding the existence of the city that bears his name there is no uncertainty. Distant enough from the sea to protect its first inhabitants from the danger of piracy, the site of Rome lay 20km upstream on the left or eastern bank of the Tiber at its lowest crossing place. This convenient ford, tucked just below an island in the river, was overlooked by a group of hills that harboured an adequate number of freshwater springs, while the surrounding countryside was suitable for tilling, grazing and hunting. The hills themselves were well wooded, fairly precipitous and defensible. In that way, the site allowed escape for early settlers from flooding and some protection against predators. This is the Rome that concerns us here, the non-grandiose Rome of the turbulent centuries when Italy consisted of a patchwork of settlements and peoples, among them the Celts in the north, the Etruscans in the centre, the Sabines next door, the Samnites along the spinal massif and the Greeks on the southern coasts. This was a time when the villagers of Rome struggled to survive, and viewed from a distance of several hundreds of years it seems one of constant conflict, as the different peoples strained for living space and the bare necessities of life.

When the hill-built city of Romulus joined the ranks of those cities that have been once, however briefly, the greatest on earth, it would become fashionable to call the rise of Rome prodigious. Cicero himself gave it an air of the miraculous when he boasted 'that Romulus had from the outset the divine inspiration to make his city the seat of a mighty empire' (*De re publica* 2.10). Yet the Rome of Cicero and the Rome of Romulus belong to two different universes, which we now know rather firmly from local archaeology. In the early days of its career nothing seemed to single out for future greatness a puny riverine settlement that lay sleeping. We should not assume, from our present vantage point in time, that this mighty empire came about by some automatic, let alone divine, process. In these obscure times Rome was allied with other Latin settlements in Latium (now Lazio), the flat land south of the river Tiber's mouth,

Marble panel (Rome, MNR Palazzo Massimo alle Terme) from a 2nd-century AD altar from Ostia dedicated to that divine couple, Venus and Mars. From bottom right to top right: the river-deity Tiberinus, the starving twins suckled by the she-wolf, the eagle of Iuppiter, the Palatine hill, the twins as hunters and shepherds and their foster father Faustulus. (Marie-Lan Nguyen)

and the seasonal battles that preoccupied these Iron Age Italic people were little more than internal squabbles over cattle herds, water rights and arable land. Rome did not flourish suddenly; nor did it simply happen. Romulus' Rome patently was not Cicero's Rome.

When, during the last two centuries of the Roman Republic, the first writers of Roman history, collectively known as the annalists, set about their job, they looked back into murkiness. Thus the most celebrated of them, Livy, whose first ten books are our single most important source for the story of Rome from its origins, mixes genuine historical material with a heap of legend, speculation and mythology, from which it is difficult to extract the truth. However, the human curse of imagination and the tendency to guess aside, these myths are of tremendous importance because they furnish us with significant clues. By comparing the written record – confused as it is – with evidence from archaeology, it is possible to reconstruct, at least in outline, the origins of Rome.

While annalistic tradition places the foundation of Rome in the year 753 BC, local archaeology tells us that in the beginning there were two separate and distinct palisaded settlements, one on the Palatine and one on the Esquiline. The Palatine was the supposed site of Romulus' city, and his thatched hut, the *casa Romuli*, was preserved there down to Livy's day as a sort of museum piece (Livy 5.53.8, Dionysios of Halikarnassos 1.79.11, Cassius Dio 48.43.4). Its postholes are still visible on the south-west corner of the Palatine, a spot Plutarch names as 'the so-called Steps of Fair Shore' (*Romulus* 20.4), but he must have meant the Scalae Caci, the Steps of Cacus. Indeed, investigations

have proved the existence of Iron Age wattle-and-post dwellings and pit burials (*a pozza* cremations) on this spacious hill at the time of the traditional foundation, and even earlier. As with the Palatine, archaeological evidence also exists for Iron Age settlement on the Esquiline. Although the inhabitants of these hilltop hamlet communities partook of the same Latial culture, which is diagnosed by the hut urn, finds from this site have their parallels at Tibur and in southern Latium, those from the Palatine being closer to the Villanovans of the Alban hills in typology. Moreover, the Esquiline trench burials (*a fossa* inhumations), which belong to the 8th and 7th centuries BC, contain grave goods that suggest an intrusion of the Sabines, a theory considered far too audacious by some scholars.

Rome's founding by Romulus is traditionally dated to 753 BC. Abandoned at birth, Romulus was believed to have been suckled by a she-wolf and raised by the wife of a simple shepherd. The sharp truth is rather more mundane as the myth is much later than this, but the date itself is plausible. This is an Etruscan sculpture known as *Lupa Capitolina*, from around 450–30 BC, with the suckling twins added in the late 15th century, possibly by Antonio Pollaiuolo. (Fields-Carré Collection)

The original migrants to the site of Rome were not primitive hunters. They were also pastoral people who had learned the art of cereal-based agriculture. Their social organization was below the institutional level of the state; for our purposes we can call it tribal. It seems likely that the easily defended hills of Rome, rising at a convenient crossing of the Tiber and with good pasture, attracted two separate bands of semi-nomadic warrior herdsmen down from the Alban and Sabine uplands. There is, therefore, some substance in the legend, as retold by Livy, that the first settlers came from Alba Longa (1.3.4), the native city of Romulus and Remus, and of the fusion between the Romans and the Sabines (1.13.6). Despite the very shadowy nature of Numa Pompilius, traditionally the second king of Rome, his name is definitely Sabine. And so Rome began as squalid clusters of herders' hovels that formed independent hamlets with communal cemeteries linked to them, which coalesced only gradually and painfully into a unified village settlement (the process contemporary Greeks knew as *synoikism*, or 'joining together'). It was all very rough, savage and ugly – indeed, hardly the sort of place anyone wrote histories about really. This, then, was the Rome into which our first Roman warrior was born.

Isola Tiberina, the island in the Tiber that divided the river's current and made it easy to ford at this point. The hills on the left bank overlooking the ford also singled out this site as one desirable for human habitation. It is said to have been called the Albula and to have received the name Tiberis from Tiberinus, a king of Alba Longa who was drowned crossing its swift, yellow waters. (Fototeca ENIT)

CHRONOLOGY OF MAJOR EVENTS

Without dates, historical discussions kindle a kind of giddy weightlessness that is the result of an inability to reason sequentially. The dates for Roman affairs before 300 BC, when the Roman list of consuls (*Fasti Consulares Populi Romani*) is secure, are those of a chronological system worked out by the Roman antiquarian Atticus, an intimate friend of Cicero, who published it in his *Liber Annalis* in 47 BC. This chronology was accepted by the greatest Roman antiquarian of all, Marcus Terentius Varro (hence the modern term 'Varronian chronology'), and was correspondingly employed thereafter by the Roman state as the official system for determining an absolute date from the supposed foundation of Rome (*ab urbe condita:* AUC). Varronian chronology dates the foundation of Rome to the year 753 BC (21 April, to be exact), the first year of the Republic to 509 BC, and the Gallic sack of the city to 390 BC.

753 BC	Traditional date for foundation of Rome by Romulus (814 BC according to Timaios).
535 BC	Carthaginian–Etruscan fleet engage Phocaeans off Alalia.
524 BC	Cumaeans defeat Etruscans of Capua.
509 BC	Traditional date for expulsion of Rome's last king, Tarquinius Superbus.
508 BC	First treaty between Carthage and Rome (according to Polybios). War with Lars Porsena of Clusium.
504 BC	Cumaeans aid Latins against Etruscan army led by Arruns, son of Lars Porsena.
499 BC	Latins defeated at Lake Regillus (496 BC according to Dionysios of Halikarnassos).
491 BC	Raid of Coriolanus.
495 BC	Tarquinius Superbus dies at Cumae.
486 BC	Rome and Latins form alliance with Hernici.
484 BC	Dedication of temple of Castor and Pollux in Rome.
479 BC	Battle of river Cremera.
474 BC	Etruscan fleet defeated by Sicilian Greeks off Cumae.
446 BC	Creation of office of quaestor (two annually elected).
431 BC	Romans defeat Aequi and Volsci on Mons Algidus.

430 BC	Rome makes eight-year truce with Aequi.
426 BC	Rome's capture and annexation of Fidenae.
423 BC	Samnites seize Capua.
421 BC	Samnites seize Cumae. Number of quaestors increased to four.
406 BC	Start of war with Veii. Romans introduce pay for military service.
403 BC	Recruitment of cavalrymen with their own horses to supplement the *equites equo publico* (who rode public horses).
396 BC	Rome's capture and annexation of Veii.
390 BC	Romans defeated at battle of the Allia. Gauls sack Rome (387 BC according to Polybios).
388 BC	Romans defeat Aequi at Bola.
381 BC	Rome extends citizenship to Tusculum.
380 BC	Rome defeats Praeneste.
378 BC	'Servian' wall begun.
363 BC	Creation of the office of praetor (one annually elected).
362 BC	Henceforth Romans annually elect six military tribunes to serve under consuls.
361 BC	Rome's capture and annexation of Ferentinum.
357 BC	Gauls raid Latium.
354 BC	Treaty between Rome and Samnite League (350 BC according to Diodoros). Rome defeats Tibur.
349 BC	Gauls and Sicilian Greeks threaten Latium by land and sea.
348 BC	Second treaty between Carthage and Rome.
343–41 BC	First Samnite War (doubted by some scholars).
340–38 BC	Latin War.
338 BC	Latin League dissolved. Roman maritime colony at Antium.

334 BC — Latin colony at Cales.

332 BC — Treaty between Rome and Molossian *condottiere* Alexander of Epeiros.

329 BC — Roman maritime colony at Terracina.

328 BC — Latin colony at Fregellae (just in Samnite territory).

326–04 BC — Second Samnite War.

327 BC — Romans introduce prorogation.

321 BC — Romans humiliated at Caudine Forks.
Romans surrender Fregellae.

ITALY BEFORE ROME

As with most regions in the Mediterranean basin, the country now known as Italy is fairly divided into barren mountain ranges with poor soil, and low-lying lush lands here and there where a coastal plain opens out, a topography that encouraged regional separatism. Around the beginning of the Italic Iron

According to the annalists, early Rome had both a Latin and Sabine element in its population. The tale begins with Romulus and his men abducting wives from their more established neighbours. This is *The Sabine Women* (1799) by Jacques-Louis David (1748–1825), showing Hersilia bringing peace between her husband, Romulus (foreground right), and her father, Tatius (foreground left), during the battle between the Romans and the Sabines. (Ancient Art & Architecture)

The Po (Latin *Padus*, Greek *Eridanus*) delta at Fratta Polesine, near Rovigo. The longest river (652km) of the Italian peninsula, the Po stretches from the Cottian Alps in the west to the Adriatic Sea in the east. It winds in a broad, fertile valley, and, as the 5th century BC dawned, a flood of Celtic immigrants (termed Gauls by the Romans) from continental Europe came over the Alps and settled here. The Romans called this region Gallia Cisalpina, 'Gaul on this side of the Alps'. (Fototeca ENIT)

Age (*c*.1000 BC) – like so many other of the chronological dividing lines in our past, this one is arbitrary – a number of regional populations can be identified and given distinct ethnic labels. They can be differentiated partly by their language, and partly by distinctive customs such as the use of characteristic artefacts, burial practices and religious cults.

The tradition of writing history only began in Rome in the 3rd century BC, in imitation of Greek historiography. What the Romans adopted, however, was not the epic military history of Herodotos and Thucydides but local history; that is to say, an account of a single city following a year-by-year chronicle format, hence called *annales* in Latin. It was an inward-looking tradition, focused entirely on the city of Rome, and though it was largely concerned with the wars of Rome, the world was viewed through Roman eyes. In this way the Romans (much like the Greeks and the Chinese) defined their own identity in terms of not being 'outsiders'. One example should suffice here. It will be remembered that Rome, in 390 BC, fell victim to the Gauls. However, the hard fact is that their progress southwards through the Italian peninsula was felt no less intensely by the Etruscans, by the Latins and by the peoples speaking related Italic languages. Naturally, this did not concern the Romans when they came to write their history.

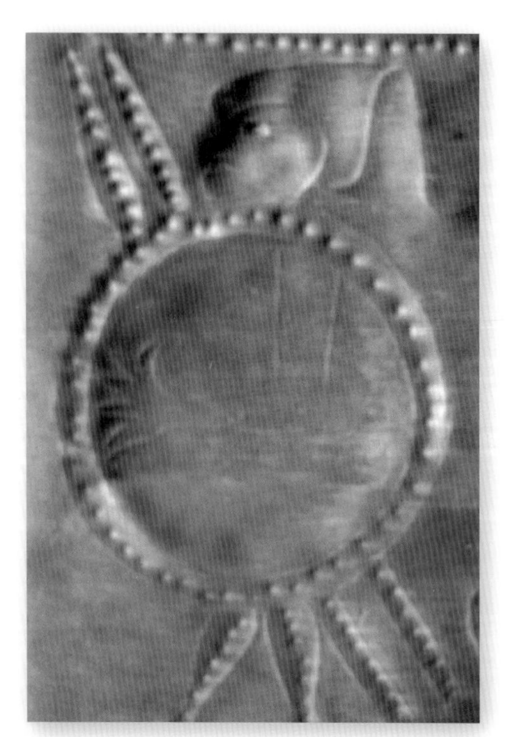

Italic votive bronze plaque, 5th century BC, showing a Venetic warrior of north-east Italy. Along with a *clipeus*, he has a helmet in the remarkably old-fashioned Villanovan style, once the hallmark of the well-dressed Italic warrior. The Veneti, who spoke a language close to Latin, had developed a very original culture, known as Atestine. However, by our period it had been subjected to numerous influences, chiefly Etruscan and Celtic. (Fields-Carré Collection)

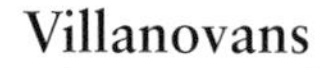

Villanovans

The term Villanovan is generally used as a reference for the Early Iron Age inhabitants of northern and central Italy, whose practice of cremation and cinerary-urn burial differentiated them from the indigenous peoples of early Italy (namely the Picentes, Umbrians and Sicels), who inhumed their dead. The Villanovans burned their dead on woodpiles, then put their ashes into biconical jars, covered them with inverted bowls or imitation helmets and buried the clay urns in pit graves with a few humble belongings.

The Villanovans are believed to have migrated into the Italian peninsula across the Alps from central Europe, the third in a wave of migrations (small populations on the move rather than invading hordes) that occurred around 1100–1000 BC. Archaeological finds suggest a close cultural connection to the new iron-using culture of central Europe. This, the Hallstatt culture, prospered from control of and trade in salt and iron. Fortunately, it requires no detailed discussion here. Suffice to say, the Villanovans spread through the Po Valley of northern Italy southward into Etruria and Latium, and Villanovan pit graves have been found as far south as Capua in Campania.

Archaeologists trace their development in two distinct periods, the Proto-Villanovan culture (1100–900 BC) and Villanovan culture proper (900–700 BC). The later period corresponds with increasing contact with coastal Greek (and Phoenician) traders and the rise of the first Etruscan city-states in what had been Villanovan settlement areas. Though Herodotos (1.94, cf. Strabo 5.2.2) believed that the Etruscans derived from the Tyrrhenians (cf. theories possibly connecting them with the Sea Peoples known as the *Trš*, or Teresh), a people that purportedly migrated by sea to Italy from Lydia (or Lemnos, if you will) 700 years or so before he lived and wrote, most modern scholars doubt this 'oriental thesis' of Etruscan origins (as did Dionysios of Halikarnassos in 1.30.1–2) and suggest that the earlier Villanovan culture gave rise to the Etruscans.

Hut-shaped cinerary urn of terracotta (Tarquinia, Museo Archeologico Nazionale). The hut urn is a model of a single-room house, and faithfully reproduces the simple, unrefined homes in which the people of the Latial culture for the most part lived. Hut urns may have symbolized the houses of the dead or perhaps had implications of social status. (Ancient Art & Architecture)

Etruscans

Etruria deserves more than a passing mention in our story. It corresponds to Tuscany in the broad sense, between the Arno and Tiber rivers, from the Apennines to the Tyrrhenian Sea, a thin but rich volcanic land whose very name recalls the people we know as the Etruscans (Greek *Turrhênoi* or *Tursênoi*, Latin *Etrusci* or *Tusci*, cf. Etruscan *Rasenna* or *Rasna*), the original creators of Italy. Possessing their own distinctive language, customs and social structure, the enigmatic Etruscans were probably not Indo-European. Although we have many of their texts, their language, which can be deciphered since it uses the Greek alphabet, has yet to be fully understood. Likewise, the question of their origins is hotly debated, an academic matter that need not detain us. What is clear is that their material culture developed out of the later stages of the Villanovan culture of central Italy as a result of increasing contact with the Greeks plying western waters, though it must be stressed that the term Villanovan does not imply a direct and definitive identification. As stated, it indicates shared cultural traits, but does not define ethnicity.

Just 16km from Rome, the Etruscan city of Veii commanded the right bank of the Tiber, with a bridgehead on the left bank at the rocky citadel of Fidenae to guard the crossing, and from it a network of routes spread out to all the other cities of south Etruria. It was in a real sense the gateway to Etruria. It was also close to Rome, and thus became its early major rival. This is a general view of Formello, Lazio, which covers part of the former city of Veii. (MM)

The Etruscan heyday was in the 6th century BC, when the Etruscans expanded at the expense of their Italic neighbours; north across the Apennine watershed and into the Po Valley, and south down in to Campania, where the Greeks had arrived before them. But the political structure that underlies this expansion remains a mystery. One of the settlements that passed under Etruscan control was Rome, where an Etruscan dynasty was installed in the closing years of the 7th century BC. The site had an obvious draw to the Etruscans, namely it was the last point before the sea where the Tiber could conveniently be crossed and so it gave access for them to Latium and southwards into Campania, with its rich soil and abundance of ore. But there were other attractions in this part of the 'golden fringe' of the Mediterranean, the most important being that essential commodity in life, used for preserving and seasoning food: salt.

To this day we say: 'a man who is worth his salt'. In early Rome, this was literally true. Our word 'salary' comes from the Latin word for salt, *sal*, and *salarium*, a payment made in salt, which linked employment, salt and soldiers, but the exact link is unclear. The elder Pliny states that 'the soldier's pay was originally salt and the word salary derives from it' (*Historia Naturalis* 31.41.89). More likely, the *salarium* was either an allowance paid to Roman soldiers for the purchase of salt or the price of having soldiers guard the supplies of this much used and highly prized commodity moving along the ancient salt route, the Via Campana, which led from the only saltpans in western central Italy. Those laying at the mouth of the Tiber on the right bank, Campus Salinarum, pass Rome's doorstep and so up the Tiber to Etruscan cities such as Clusium and Perusia. In later times, the road leading north-west out of Rome into the Sabine interior and then onto Umbria was known as the Via Salaria, the Salt Road.

Like the Greeks, the Etruscans were bound together by a common language and culture but were politically organized into a loose confederation of largely autonomous cities. The 6th century BC also saw the Etruscans with a considerable war fleet, and in 535 BC they joined the Carthaginians in driving the Phocaeans from the seas at Alalia. In 524 BC the Etruscans of Capua attacked Cumae, but were defeated. When the Etruscans of Clusium had to face the Latin alliance against them 20 years later, they found Cumaeans in the ranks of their enemies. There is no evidence for Etruscan military or political unity against a common enemy (cf. the fate of Veii), but it does appear that interstate conflict in Etruria was common. Thus, the question remains whether or not such a socio-political set-up could have supported their distant conquests into the Po Valley and Campania.

The Etruscans, the most cultured of the Italic races, extended their influence northwards nearly to the Alps and southwards over Campania. From Etruria Rome borrowed many ideas and concepts, including those for military and political purposes, and at least two of its seven kings were said to be Etruscan. This is a horseshoe-shaped sandstone funerary stele (Bologna, Museo Civico Archeologico, inv. 164) from Felsina, around 400 BC, showing an Etruscan horseman encountering a naked Gaulish warrior. (Ancient Art & Architecture)

Woodcut engraving by John Reinhard Weguelin (1849–1927) showing Castor, the breaker of horses, and Pollux, who was good with his fists, fighting at Lake Regillus. A popular Roman legend had the Heavenly Twins, appearing as two young horsemen, help in gaining victory for the Romans. (Reproduced from Thomas Babington Macaulay, *Lays of Ancient Rome*, 1881 edition)

Latins

The Latins (*Latini*), who settled in the open country south of the lower reaches of the Tiber (hence its name of Latium), are the ethno-cultural group to which the Romans mainly belong. Originally, in the Early Iron Age, these people of west-central Italy consisted of a group of 30 communities, of which, in the beginning, Rome was only one. They spoke the same language, Latin, a subgroup of the Latino-Faliscan branch of the Italic language family, and each year they gathered to celebrate the festival of Iuppiter Latiaris on the Alban Mount (Monte Cavo), the highest point in Latium (Dionysios of Halikarnassos 4.49.3). Archaeology has demonstrated that they also had some distinctive artefacts and burial practices, such as the use of the hut urn for cremated remains. By the 7th century BC (and possibly much earlier) the Latin communities were grouped into a confederation for sacral and religious purposes, and by the following century this confederation had taken on the form of a political and military league.

Sabines

The Sabines (*Sabini*) were an Italic people that lived in the central section of the Apennines, the formidable mountain chain which forms the spine of the Italian peninsula. They also inhabited Latium north of the Anio before the founding of Rome. Their now-extinct language belonged to the Osco-Umbrian branch (formerly Sabellic) of the Italic language family and contained some words shared with other branch members, such as Oscan and Umbrian, as well as Latin. The Sabines, apparently, were the religious folk par excellence. According to the pioneer scholar of the Latin language and Roman institutions Varro (*Lingua Latinae* 5.73), it was from his pious Sabine compatriots that the Romans obtained many of their divinities, and traditionally it was the Sabine king, Numa Pompilius, who fashioned the religion of early Rome.

Monte Terminillo (2,217m), a massif in the Monti Reatini, part of the Abruzzese Apennines. It is located some 20km from Rieti (ancient Reate), once a major Sabine settlement astride the Via Salaria, and 100km north-east of Rome. This limestone range was not rich with ores or lodes, and, being landlocked, enjoyed no connection with the sea. Therefore, when not preying on their neighbours, the inhabitants were obliged to scratch out their frugal living directly from the rocky soil. (Fototeca ENIT)

After the battle of Regillus, the Heavenly Twins' cult was brought to Rome, a temple being dedicated to them in 484 BC (Livy 2.42.5). The three surviving columns of the temple of Castor and Pollux (seen here on the left) have been a landmark through the centuries, but it stands today in the form given to it at the end of the 1st century AD. The earlier temple is still a matter of mystery. (Fototeca ENIT)

Oscans

In the central and southern section of the Apennines, most of the Italic peoples spoke the now-extinct language known as Oscan, which belongs to the Osco-Umbrian branch of the Italic language family. As such, the Oscan tongue was closely related to Latin, but had some distinctive characteristics. The Oscan speakers themselves were divided into various groupings, the most important of which were the Samnites, who inhabited the mountainous region east of Rome down to the area behind Campania. At the time of their long, hard wars with the Romans, the Samnites banded themselves into a loose confederation (called *civitas Samnitium*, or Samnite League, by Livy) consisting of four distinct tribal groupings, each with its own territory: the Carricini (sometimes referred to as the Caraceni), Caudini, Hirpini and Pentri, to whom we should probably add the Frentani. But these Oscan groups often formed new tribal configurations. In the late 5th century BC a new Oscan-speaking people, the Lucanians (*Lucani*), emerged (perhaps a southern offshoot from the Samnites), and in the middle of the following century another Oscan-speaking people, the Bruttians (*Bruttii*), broke away from the Lucanians in the toe of Italy.

The instability of the diverse Oscan-speaking peoples was probably a result of population pressure. We have no demographic records, but it seems clear from archaeological data that all over Italy the population expanded at the turn of the 3rd century BC, driving rustic communities to come to blows over land. Good arable land was in particularly short supply in the upland valleys of the Apennines, which were rough and stony if picturesque and

Right: Disc-and-stud helmet (Bologna, Museo Civico Archeologico) from the Necropoli sotto la Rocca-Lippi la Tomba Principesca N.85, 7th century BC. This helmet is made of a wickerwork cap reinforced with bronze discs, the gaps between these discs being filled with bronze studs.

mountainous, and in the course of the 5th and 4th centuries BC the coastal settlements, many of them founded by Greeks, found themselves exposed to the menace of the highlanders. It seems warriors were the only crop that the Samnites grew naturally on their thin, stony soil. Years of scrambling up and down scrubby mountainsides had made their bodies immensely strong, while the harsh environment of the Apennines fostered the skills of formidable fighters.

And so it was by force of arms that they seized Greek Cumae and Etruscan Capua, eventually merging with the existing inhabitants of Campania to give rise to the Campanians (*Campani*). In the meantime, the Lucanians overran Poseidonia, renaming it Paestum but maintaining the socio-political institutions set up by the Greek colonists, and attacked other Greek cities scattered along the south-eastern seaboard. These speakers of Oscan thus imposed their language upon all of southern Italy except in the heel and in those coastal communities that remained under Greek control.

Below: Four bronze discs for a similar helmet from the Sepolcreto Benacci Tomb 494, dated from the end of the 8th century BC. This pattern of helmet is shown being worn by five of the figures on the Certosa *situla*, also from Bologna. (Fields-Carré Collection)

Bone-poor, skilled only in manual toil and weapon handling as these people were, the Oscan military ethic encouraged wars of conquest, but in these fat lands the highlanders had established something like an ascendancy that abjured the memory of their warrior fathers. So later the erstwhile conquerors, who now formed the local aristocracy, readily became spoil for the parent stock they had left behind in the highlands. No doubt there was retaliation, but in violent exchanges of this kind the advantage lay with the men of the stony mountains, who were much tougher and more tenacious. Eventually, this would allow the Romans to exploit the worsening situation and support the Campanians against their mountain kinsmen, an action that was to provoke the First Samnite War (343–41 BC). Campania was a productive and populous region, and neither side could afford to let the other get control of it: for the Romans it meant a good source of grain, as Rome was susceptible to local shortages (e.g. Livy 2.52.1, 4.52.4–6).

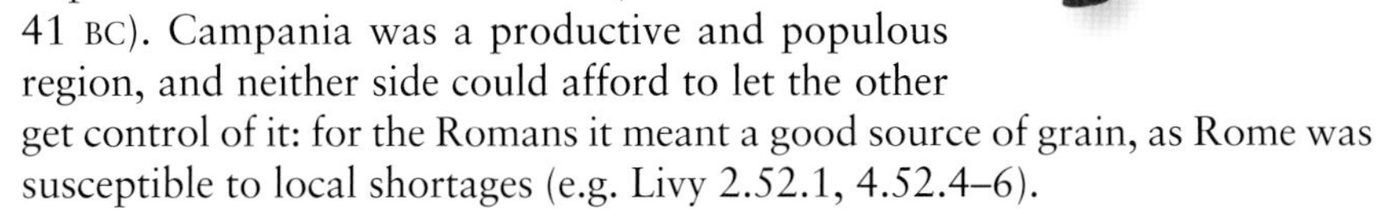

Bronze horseman (London, British Museum, inv. GR.1904.7-3.1) from Grumentum, near Metapontion (*c.*550 BC). The rider wears a Corinthian helmet (originally with a transverse crest) and surely represents an aristocratic hoplite riding his horse to battle. The *equites* of Rome would have been of this stamp. (Fields-Carré Collection)

Greeks

Beginning in the 8th century BC, Greeks took to their agile boats and began to plant colonies along the Italian seaboard. The earliest, on Pithekoussai, what is now Ischia in the Bay of Naples, was founded initially by Greeks from Euboia as an offshore haven for Greek merchants and carriers on the coastal voyage to willing Etruscan customers. When the colonists felt sure of their surroundings they established a second colony at Cumae on the opposite Italian coast, just north of the Bay of Naples. The traders were soon followed by settlers.

From the late 8th century BC other Greek settlements were founded on the fertile coastal plains of southern Italy and Sicily (and beyond) so as to relieve population pressures back home. What is more, Greek soil was poor, rocky and waterless, meaning that perhaps no more than 20 per cent of the total surface area of the peninsula could be cultivated, so these colonies were soon to become sources of that threefold triumph of wheat, olive and wine for the mother cities (New World crops – potatoes, tomatoes, tobacco, maize, etc. – were unknown in antiquity). Nonetheless, these colonies, almost as Hellenic as their mother cities in old Greece, were politically autonomous from them, though they normally retained close cultural and sentimental links. Colonies were generally established on easily defended sites such as steep acropolises, small offshore islands or promontories of the mainland, and usually in a location where the indigenous population did not pose a major threat to Greek settlement. Our immediate concern, however, is the development of infantry shock tactics by the Greeks. But of this, more anon.

EARLY ROMAN WARFARE

There is a remarkable contrast between Rome as imagined by us today and the Rome inhabited by Romulus, the half-god, half-human son of Mars and Rhea Silvia. The prevailing impression left by the former vison is one of ease, order, peace and sumptuousness, while the reality of the latter is best associated not with peace and prosperity but poverty, squalor and bloodshed. No matter how it is dressed up, the fact is that this was a desperate and unappealing place, and the original Romans a race of rude roughnecks.

As we fully appreciate, among the legacies of the Romans bequeathed to modern man is the fully developed practice of war. Yet in the very beginning the Roman way of war was little different from that practised by other Italic peoples. In part this was because no single tribe was superior in military technology, either in the sense of the manufacture of arms or of military tactics. And not merely in technology either, for in economic and social development Rome and its neighbours were all closely placed. Existing in societies centred on war, they met quite frequently on the field of conflict, in shifting alliances and hostilities, and any technical developments would quickly have been copied or shared. The armament industry is apt to be international, and a good weapon, or a good bit of protective gear, will travel quickly once its advantages are appreciated. Likewise, tactics are very much a transferable skill.

CLAN CHIEFTAIN

Local wars and vicious raids for booty, indistinguishable in the eyes of Romulus' world, were organized affairs often involving a clan, managed by the clan chieftain, or they could be larger in scale, involving a number of clans acting cooperatively. Invariably there was an expeditionary leader. One of his functions was to act as the paramount chief and keep the restless clans together. His other, equally important, function was to carefully divide the booty derived from pillage and plunder, 'so no one', explains Odysseus, 'not on my account, / would go deprived of his fair share of spoils' (*Odyssey* 9.48–49 Fagles). Movable property was thus continually changing its owner, according to the victorious sword. This filled the hands of the victor with riches, and enabled him to gratify his armed followers, on whose strong arms his status rested. They risked their lives on his command in exchange for a share in the booty. The greater the plunder, the more his followers loved him. 'The shepherd of the people' is a Homeric commonplace. The sword-bearing chieftain thus gave protection at home and plunder in war, and one might here quote Goethe's philosophic axiom: 'He who is no warrior can be no shepherd.'

Naturally, a chieftain who failed to provide for his followers would lose them, and with them all the power and status they conferred on him. So booty fuelled these clans and their warrior bands, and such a lifestyle dictated an expansionist policy towards one's neighbours, since, in order to distribute wealth to his followers, a chieftain first had to accumulate it. But that was only part of the story. Seeking glory in combat and exerting a great fascination over his contemporaries, personal courage was obviously very important to this aristocratic warrior, and the bearing of arms, especially a long-bladed sword for slashing, may have been regarded as a potent symbol, of both free manhood and of power and wealth. Clan chieftains rose and fell by the casual brutality of the sword, and on some occasions single combats (probably fought to the death) could be formally arranged with the opposition. For these proud men there was something correct and consecrated about a flat field, a fair fight, no respite and a fair death.

They were resplendent in shining helmets, pectorals and greaves, which were fashioned from beaten bronze often beautified with embossing. Their swords were of the superb antennae type, so named after the cast-bronze tang ornamented with spiral horns. The two-edged blade, invariably of bronze (iron being comparatively rare), was designed mainly for cutting, but could also be used to thrust or jab. It was above all the sword, and the ability to use it, that constituted the chieftain's *insignia dignitatis* ('esteemed badge/mark'), and no other weapon more clearly proclaimed his authority in society or prowess in combat. Swordplay not only allowed a chieftain to display his courage but also his individuality. Here we have an example of such a clan chieftain, a man whose business was fighting. He grips an ash-wood spear and a long sword hangs at his side. On his head sits comfortably a splendid example of a Villanovan helmet, while his armour is based upon the elaborate poncho-type cuirass discovered at Narce (Tomb 43) in Etruria .

After precious metals, cattle were the most sought-after form of spoil because of their value as a measure of wealth and status. Here we should note that coinage was a very late arrival in Rome, and that *pecunia*, the Latin word for money, originally meant 'cattle' (cf. *pecus*). This is a bronze bull figurine (Vienna, Naturhistorisches Museum), Celtic Late Hallstatt, 5th century BC. (Werner Forman Archive)

With Rome's affairs confined to Rome itself, as it were, its wars appear to have been organized around raid, ambuscade and cattle rustling, with perhaps the occasional pitched battle between armies. The first three had yet to become the definitive form of war-making, where the destruction of the enemy is the goal. Pitched battles were fought by little more than warbands formed by a warrior aristocrat, his kinsmen, friends and clients, much like the clan gathering of the Fabii with its 'three-hundred and six clansmen and companions' (Livy 2.49.4) who proudly marched to battle against Veii, Rome's Etruscan neighbour just across the Tiber, and tragically died fighting at the river Cremera. We cannot know the size and maintenance of such forces for sure from the evidence available to us, but it is unlikely that numbers were large. Though friends, neighbours and clients served to extend the natural limits of a clan, these warbands could not have exceeded a few hundred men at most, and in most cases numbered far less, because of the economic and logistical constraints inherent in the subsistence-type economy (based on cultivation and animal husbandry, augmented by hunting) over which Romulus' village presided.

Volumnia before Coriolanus, by Gillis van Valckenborch (*c.*1570–1622). It is said that the young Coriolanus won his spurs at Lake Regillus. It is also said that later in his life he starred in his own legendary tragedy. According to the story, Coriolanus, now a defector at the head of Volscian raiders, stormed up to the gates of Rome, only to be turned away by the supplications of Roman matrons, including his mother (Livy's Veturia, Shakespeare's Volumnia) and wife. (Rafael Valls Gallery, London, UK / Bridgeman Art Library)

These small numbers did not detract from their effectiveness in the field, however. Even if Livy's rather exact figure cannot be accepted at face value (some would argue it has been distorted for poetic gain), and raw numbers by themselves can be rather misleading, a raiding force of 306 panoplied warriors prepared for war, as observed from the receiving end, could cause considerable damage and terror, and would be fast, versatile and predatory. Moreover, as well as maximizing the benefit of surprise, a smaller force also minimized the risk of casualties by not seeking involvement in pitched battle. In short, military matters during this pre-urban period were on a very modest and personal basis, with the clan chieftain fighting for personal glory, his retinue of armed followers out of loyalty to him and, of course, the prospect of having that loyalty rewarded with portable loot. This single fact suggests that large-scale larceny was inseparable from small-scale warfare.

Clan warfare

So raiding and ransacking the neighbours was a normal part of early Roman warfare, and the chieftains of this period probably arose from among the 'big men' common to 'warband cultures', restless and charismatic types who made good mainly through the redistribution of surplus wealth that warlike success could bring, fighting as individuals, relying only on their own pure courage and the strength of their hard bronze arms. That was certainly the kind of fighting that was bound to make the strongest men preeminent. However, Rome throughout this remote period lies almost outside recorded history and such distinguished deeds went unwritten. For all that, it is likely that the adventures of these valorous and generous warriors were passed on by bards who, in doing so, embellished details concerning the real history of their subjects. And so it takes little imagination for us to equate our clan chieftain, who looked to his bard to immortalize him in verse, with the brigand boss who seeks glory and gold in simple and unadorned predation.

Destitute as they are of historical credibility, many of the heroic tales of Rome's early history recorded by Livy (books 1–3) were not entirely a figment of his literary imagination but in fact may have had their origins in the panegyric poetry, oral in its presentation and transmission, composed to celebrate the hawkish achievements of these clans and their chieftains during this turbulent time of borderland forays. According to Cicero (*Brutus* 75, *Tusculanae disputationes* 1.3, 4.3), a near contemporary of Livy, ballads telling of the feats of olden days were once sung as a popular form of entertainment. Even so, in the course of long-term transmission and constant

The warrior of Capestrano (Chieti, Museo Nazionale di Antichità delgi Abruzzi e del Molise), dated to the second half of the 6th century BC. This limestone statue, 2.09m tall, was found in Capestrano in the Apennines and represents an Italic highlander in full battle gear. He wears disc armour held on by a leather harness. He is armed with a sword, which is slung across his chest, an axe and two javelins with throwing loops. His throat is protected by a throat guard, and his broad-brimmed hat almost resembles a sombrero. (ph. Giovanni Lattanzi / inabruzzo.it)

reinterpretation, such tales are likely to have been transformed into a potpourri of wild nonsense mixed with sober fact. After all, boasting about warlike deeds was the chief job of inspired bards.

So the story of infant Rome appears to be a confection of fluid oral traditions, confused folk memories, hoary myths, dubious romantic fiction, idle hearsay and unblushing lies. For the emperor Caligula Livy was 'a wordy and inaccurate historian' (Suetonius *Caius* 34.2). The *obiter dicta* of an immature iconoclast perhaps, but who are we to argue with an emperor of Rome? By our own standards the patriotic Livy may be a rotten historian, the historical substance of his accounts falling under the shadow of uncertainty cast by the nonexistent documentary record of Rome in its beginnings, but he was certainly a skilled journalist who liked nothing better than a good yarn. Written at a time when Rome itself was still shocked and riven by anarchy, in many respects these initial books of his great saga hark back to the so-called golden years of Rome and allowed the Romans of his day to wallow in their own history and traditions.

However, even if we must suspect that Livy indulges in a measure of bardic licence, his *History of Rome* is not a flat-footed lament for departed glory. Contemplation of Rome's past may offer more than simply an escape from the present, it may also show how to recover a former excellence and a return to the good old days and to innocence, free from the various political and social ills that plagued their society. We ourselves tend to remember the best about our ancestors and forget what in any way diminished them. The Romans even more so.

B

CLAN WARRIOR

Uniformity was never a characteristic of any warband, and the quality and quantity of weapons and equipment would vary widely, ranging from abundant to minimal. Because of archaeology, we can say with some confidence that these Roman warriors of fewer means, the military backbone of warbands, were without armour and almost certainly armed with a shield for protection and a spear for thrusting, with perhaps hand weapons such as dagger or axe for close-quarters battle. Evidence from grave goods reveals that the sword, the weapon that should be associated with the wealthier or more successful members of a clan, was the least common of principal hand weapons. In contrast, the spear and shield were plentiful, being made largely of wood, which was cheap and readily available to clansmen, who, after all, were free men who normally farmed and herded as clients of a chieftain to whom they owed allegiance. Therefore, if a clansman's complete war gear consisted of two spears and shield, then we can infer that one of the spears was thrown as a missile weapon, while the other was retained for use as a stabbing weapon once the opposing sides had closed on each other and become locked in combat.

Against this array of offensive weaponry, the clansman entrusted his safety first and foremost to his shield, which was in all likelihood the Italic *scutum* with its signature long central wooden spine, metal boss-plate and single handgrip. With the exception of all but a few of the wealthiest warriors, body armour was not worn and the existence of metal helmets rare. Obviously no one who went to war would feel entirely safe without one, and no one would pass up the chance of grabbing one if he possible could. They were an obvious target of looters after battle. Every clansman had a foundation in simple skills, such as sewing, repairing equipment, replacing rivets and the like, which they used to enhance the protection of the gear they scavenged, looted or otherwise acquired.

Though a variety of different helmet patterns are known – cap, conical, bell, pot, disc-and-stud, broad-brimmed – simple skullcaps of *cuir bouilli*, fitting snug and tight, were used at the least, and anything that would protect the head from the blows of the enemy could have been pressed into service, such as wickerwork reinforced with discs or plates of bronze. Each man would bring whatever he could afford or could scrounge on an individual basis. For the most part, however, the only things that prevented a clansman's sudden death or serious injury in the hurly burly of battle were his *scutum*, large enough to screen a crouching man, his own martial prowess and his physical strength and agility.

Solid cast-bronze statuette of Italic warrior (Bologna, Museo Civico Archeologico, inv. IT 1281) found in Reggio nell' Emilia and dated to the first half of 6th century BC. Though the warrior appears naked, and may indeed represent a deity, he is clearly depicted wearing a Villanovan helmet. Votive figurines such as this one show us how the Villanovan helmet, with its exaggerated crest, was worn. (Fields-Carré Collection)

The Rome of Romulus' day was anything but glorious. It was deliberate and rapacious in its habits, an emerging society jostling for space on the Italic stage. In particular, it was vulnerable to the highlanders living in the mountains to the east. Rome tried to intimidate these mountaineers by frequent raids, which only encouraged them to retaliate. The predatory and opportunistic behaviour of these early Romans is ideally illustrated by the flurry of raids and counter-raids conducted against these local highlanders (the Sabines, the Aequi, the Hernici and the Volsci), which Livy conveniently labels as frequent instances of '*nec certa pax nec bellum fuit*' – 'neither assured peace nor open war' (2.21.1, cf. 48.5). This was an ugly war of ambuscades, with surprise attacks followed by equally rapid retreats, artful deceits followed by face-saving compromises – a hide-and-seek game. In this, the age of the clan, loot mattered as much as loyalty to chieftain or to Rome; it was the key to courage in combat. Pillage was not simply the inevitable and distasteful consequence of war, but the very substance of it.

Battles fought, chieftains slain, villages pillaged, strongholds besieged and fields burned – these are the ingredients of Livy's stew (e.g. 2.23.5, 30.11, 33.7, 39.5, 50.4, 51.5, 62.1). Clearly he is putting ample flesh on the barest of historical bones. And so, much like that celebrated line has it in the great Irish tale *Táin Bó Cúailnge*: 'Men are slain, women stolen, cattle lifted.' The subject matter of war was the villagers and the victuals, the cows and the crops, and what we are witnessing here is tribal as opposed to state warfare, where there are battles but no campaigns, tactics but no strategy, and in the eddying fortunes of warband conflict the purpose of leadership is generally to inspire.

Yet we should resist the conclusion that these intertribal hostilities arose from economic motives. To the modern mind this perhaps seems odd. Of course, all these warring tribes, the Romans included, assumed victory would bring some well-deserved material benefits to the victors, such as the lifting of cattle herds, but the underlying cause was to avenge wrongs and to uphold honour, not to look beyond battle to the fruits of victory. Warfare was conducted as though it were an internecine blood feud, with one tribe seeking to exact justice or vengeance from another. To this feature we can add by saying that much of the fighting of the period was undertaken more for the love of fighting and from a spirit of adventure than in order to achieve definite economic (and political) results. Indeed, in economic terms predation was a totally unproductive activity as it merely moved to and fro the fruits of other tribes' labours. Indeed, raiding the neighbours and carrying off their women and cattle was only workable if the neighbours would and could counter-raid to recycle the booty. For the Romans, war had yet to become a complicated, responsible, prosaic business; that is to say, it was not yet the stuff of ambition and fought for the love of power.

Early Roman warfare, a virtually unbroken continuum of thatch burning and cattle lifting, was not an exclusively elite pastime. In such a world, however, the heroic values of prowess, courage, generosity, an insatiable desire for fame and an overwhelming fear of disgrace, acted to strengthen the social and cultural underpinnings of a nascent Rome struggling to define itself and gain a foothold against its powerful neighbours, setting a standard of conduct and behaviour by which all its warriors were to be judged. This created a predatory form of society with power becoming centralized in the strong arms of a select few, men more drawn to the terrible swift sword than the trusty plodding ploughshare.

Bronze antennae sword (Bologna, Museo Civico Archeologico) from the Sepolcreto Benacci-Caprara Tomb 39, dated from the end of the 8th century BC. Most swords of this type have long points, which allowed them to be used for thrusting as well as slashing. This example, however, has a slightly curved sabre-like blade, which suggests a preference for slashing rather than thrusting. (Fields-Carré Collection)

City-state warfare

As we recall, during the age of kings there was no appreciable difference between the Romans and their neighbours in military matters. Rome's population and territory were not large, and their neighbours, which it raided and was in turn raided by on a regular basis, were often barely a day's march from its own gates. The raiders swept down and passed on. On the other hand, if a neighbouring community was destroyed, usually by denying it its means of life via vandalism and theft, its fields were acquired by Rome and the conquered villagers often deported to the fledging city. This apparently happened to Ficana when the king, Ancus Marcius, took this riparian settlement sitting just 11 Roman miles (16.25km) down the Tiber (Livy 1.33.3, Dionysios of Halikarnassos 3.38.3). Whatever the worth of such annalistic traditions, these threatened hamlets and villages could often gain security by yielding before an attack and the population might become the clients of the king or some clan.

Kings were often driven by the desire to conquer and, as a result, internal disputes gave way to confrontations on a more extraneous scale. All the kings (except Tarquinius Superbus) increased the size, both in area and population, of Rome (Livy 2.1.2). Yet 'membership' of Rome was not simply a status that one did or did not possess. It was an aggregate of rights, duties and honours, which could be acquired separately and conferred by instalments. Those populations seen as ethnically and linguistically close to Rome were eventually admitted to full Roman citizenship. To those less close to Rome, a sort of half-citizenship, under Latin law, was sometimes offered, but they were liable for Roman military service. On the sites of former settlements, or on land not yet settled, garrison colonies were planted, either Roman (*coloniae civium Romanorum*), in which case they were peopled with Roman citizens, or Latin (*coloniae Latinae*), with some autonomy but fewer rights than the former. Another possible status was that of ally, *socus*, with or without treaties granting

Funerary art from Paestum (Andriuolo Tomb 114, c.330 BC). This battle scene depicts two hoplite phalanxes about to clash head on. The one on the right is believed to be the winning side, as it is led by a figure, heroically nude and in the act of thrusting with his spear, perhaps to be identified with the pan-Italic Mars. Note the individual blazons on the hoplite shields. (Fields-Carré Collection)

equal rights but still with an obligation to supply Rome with military manpower, which was to be a vital factor in Rome's ability to wage war continuously.

And so Rome began its long career of conquest and control through a common-sense policy of incorporation, bit by bit absorbing all its nearby rivals and gradually growing, and as it grew so too did the scale of its conflicts. Ambuscades, raiding, plundering and slaughter, naturally, were still common, but there was a gradual shift to pitched battles, which required far greater military organization and resources. The choice of military response to win or protect territory was now to be a civic matter, and the rough warbands and their heroic chieftains were replaced by a wider levy of all those adult males who could provide themselves with the appropriate war gear with which to fight.

Though it is said (e.g. Diodoros 23.2.1) that the phalanx formation came by way of the Etruscans, a crude and transparent argument that plays up the native Italic tradition, in all probability this change in tactics owes its origins to the Greek cities that fringed the coasts of southern Italy. Inspired either by the Etruscans or the Greeks, the adoption of mass fighting in tight formations, and of mustering a militia army organized around a phalanx composed of citizens wealthy enough to outfit themselves with the full panoply of an armoured spearman, was radical. And so by the time Rome was no longer the hilltop village on the banks of the Tiber, the Roman way of war had changed from an agglomeration of numerous single combats to become an adaptation of hoplite warfare and the hoplite ideology of the decisive battle. Gone was tribal conflict, where each warrior could feel a personal commitment and justification in the face of the enemy. Individualism had been ceded to collectivism, mobility traded for protection. 'Depersonalized' warfare had arrived.

LEVYING

As far as we can tell, the earliest government of Rome comprised a king (*rex*) with military, religious and political power (*imperium*), a consultative council (*senatus*) of elders (*patres*) drawn from the chieftains of the ruling clans (*gentes*) and a consultative assembly (*comitia curiata*) constituted on a federal basis from the various hamlets (*curiae*). Since an early Roman community required a king of vigorous adult years who could protect it and give it his personal leadership in war, these kings, at least two of which are said to have been adventurers of Etruscan origin, were not hereditary monarchs, each king being inaugurated by consent of the gods and acclaimed by the people. Also, while kings fought and kings fell, the *senatus*, the senate, lived on.

Without failing prey to excessive trust in the annalistic tradition, it seems clear that Rome was eventually drawn into the Etruscan orbit, with that becoming part of an urbanized region. With urban planning, drainage and street systems came the related development of organized social, administrative and political functions and structures. At this time, around 625 BC, Rome was politically unified by the creation of a single, central marketplace, the *forum*

Stele from Tarquinia (Monterozzi Necropolis, Tomb 89), 7th century BC. Whereas clansmen were best equipped for and accustomed to cattle raids and skirmishes, hoplites were armoured spearmen who fought shoulder to shoulder in a phalanx formation. These citizen-soldiers were now protected by helmet, corselet and greaves, all of bronze, and wielded a long spear and large shield. As well as thrusting with their spears, hoplites pushed with their shields. (Fields-Carré Collection)

Campanian breastplate of bronze, from around 375–25 BC, moulded into the form of the muscles of the torso. Like others of its kind, this armour is far too small for the musculature to fit the torso of a normal man. Thus the naval and the pectoral muscles were never intended to cover the corresponding parts of the wearer, and this breastplate (along with its matching backplate) would have been worn above an Oscan belt. (© Board of Trustees of the Armouries (II. 197 - TR.2001.185)

Romanum (the Roman Forum), and located there were certain communal buildings such as the shrine of Vesta and a palace, the Regia (later the seat of the *pontifex maximus*). In addition, the hamlet system was dissolved by the creation of three tribes, Ramnes, Tities and Luceres – all Etruscan names and thus ostensibly the result of the influence of Rome's powerful northern neighbours – which were not based on residence or ethnic origin.

With the transition from pastoral primitivism to urban sophistication, the inhabitants of Rome therefore became one people: they were Roman citizens. All the same, kingship is the simplest form of government and also the most easily corrupted, since its proper functioning depends upon the character and abilities of a single individual. The end of a monarchal system of government can ensue merely from a tenure by one person unfit to occupy the office, and perhaps this was the simple case with Rome when the Romans replaced the king with two annually elected magistrates who shared equal power. Livy (2.1.2–6) himself acknowledges the rule of the first kings as a necessary stage in the socialization that prepared Rome's motley population for the republican self-rule that followed, and at the same time makes clear his own view that the freedom of self-governing people is superior to monarchy.

Clan gathering

Though regard for accuracy was not to be allowed to interfere with a cracking good yarn of bygone Rome, annalistic tradition does emphasize the importance of the clan rather than that of the civic community. As we know, Livy speaks of *gentes* (clans) formed by *cognati* (blood relations) – called *sodales* and *clientis*. The latter was a circle of people gathering around a powerful patron, namely the clan chieftain, who offered him services in return for protection, while *sodales* were the chieftain's blood

The story goes that Numa Pompilius saw a shield fall from the sky and, having retrieved it, ordered another 11 copies made so as to hide the divine original, it having been prophesied that Rome would endure as long as this shield remained there. These bronze shields, *ancilia*, resembled a rough figure-of-eight. (Photo courtesy of Nick Sekunda)

relations and also his companions and his attendants. So warbands would have included many men related to each other by blood, marriage and other, 'fictive', ties. As Homer justly writes, 'Brothers a man can trust to fight beside him, true, / no matter what deadly blood-feud rages on' (*Odyssey* 16.109–10 Fagles). It was important to stand alongside people whom you knew and trusted.

The warrior of tribal Rome took a broader view of war when he was summoned from his herds to slay the marauder or to pillage his neighbour. Though he did not formulate his conviction, he knew that the ultimate aim of armed force was for the good of the community. In other words, Roman armies were held together by threads of social obligation.

In times of conflict a clan chieftain would collect his own family members and call on those of his relatives and clients. These warrior-farmers would form a clan-based warband and take rations with them, from household stores, for the two or three days that the raid might last. Collectively they might make one or two such raids per season, so it would not impose a great strain on farming manpower or food stocks. With a national levy, namely a specially mustered host led by the king in person, the scale of ambition shifts from the small and limited to the large and (relatively) unlimited. With this size of force it is possible to fight pitched battles – if that is the word – and pillage whole regions rather than merely hamlets and homesteads.

The composition of armies may have varied according to their function (e.g. cross-border raiding or home defence), but each would normally have consisted of an agglomeration of warbands fighting under the command, and loyal to, individual clan chieftains, while the army as a whole could have been under the patriarchal aegis of the king himself. As the leader of his people, the king had a solemn obligation to protect them and their property against the depredations of neighbours and to lead them on expeditions of plunder and conquest. For the conduct of such warfare, the king was undoubtedly empowered to summon the clan chieftains and their followers to a mustering. These clan gatherings were disbanded at the end of a military operation and the clansmen went back to work on the land, to be summoned again when the need arose. The king was essentially the leader of many clan-based warbands.

Citizen muster

Livy (1.42.4–43.8) and the Greek historian Dionysios of Halikarnassos (4.15.6–19.4), both rhetoricians and both writing in Rome under the reign of Augustus (27 BC – AD 14), attribute a major reform of Rome's socio-political and military organization to the popular king Servius Tullius (r. *c.*579 – *c.*534 BC). His first consideration was the creation of a citizen army, and the most important point was to induce the citizens to adequately arm themselves. So a census of all adult male citizens recorded the value of their property and divided them accordingly into five economic classes. Whether or not Servius actually existed, archaeological data and comparative studies do suggest that the Romans adopted hoplite panoply around this time, so the annalistic tradition, as seen by Livy and his Greek contemporary Dionysios of Halikarnassos, may be broadly accurate. Therefore, though the reform scarcely sprang fully fledged from the brain of Servius, it must be remembered that many of its principles – rather than its details, which were elaborated only after years of gradual development – belong to this period.

In Livy (1.43.1–7) the Servian class I essentially fought with a hoplite panoply, each man equipping himself with helmet, corselet and greaves, all of bronze, together with the *clipeus*, that is to say, the *aspis* (large round shield) carried by Greek hoplites. The spear and sword were the main weapons used. Men of class II equipped themselves similarly, but were not expected to provide a corselet, while those of class III could omit the greaves as well. However, to balance the absence of body armour, classes II and III used the oval *scutum* instead of the round *clipeus*. This was a body shield, Italic in origin. Finally, classes IV and V were armed as skirmishers, the last class perhaps carrying nothing more than a sling. Generally speaking, the historical process is a rather messy business and often works itself out in total defiance of neat schemes. Thus, it has been suggested there were in fact two stages of development here, first a single undifferentiated class, or *classis*, of those who possessed the minimum qualification to serve as hoplites with all the rest named *infra classem*, 'below the class', with a fivefold subdivision coming much later. This hypothesis would certainly better reflect a period when the art of war was still in development.

More important was the subdivision of these five classes into *centuriae*, or centuries: in each class half the *centuriae* were made up of elder men (*seniores*, those aged 47–60), obviously more suitable for garrison duty, and half of younger men (*iuniores*, those aged 17–46). The number of *centuriae* in each class were unequal in number, as the state naturally drew more heavily upon the well-equipped wealthier men than on those lower down

CITIZEN-SOLDIER, CLASS I

Servian class I citizen-soldiers fought essentially with hoplite panoply, each citizen equipping himself with helmet, two-piece corselet and greaves, all of bronze (though later linen and composite corselets would be usual). He also carried the *clipeus*, a bowl-shaped shield, approximately 90cm in diameter and clamped to the left arm. There is a superb example of a *clipeus* in the Museo Gregoriano at the Vatican. This shield, which probably comes from an Etruscan tomb of the 4th century BC, has survived sufficiently intact to permit a complete reconstruction with a good deal of confidence (Connolly 1998: p. 53).

Built on a wooden core, this shield was faced with an extremely thin layer of stressed bronze and backed by a leather lining. The core was usually crafted from flexible wood such as poplar or willow. Because of its great weight the shield was carried by an arrangement of two handles, with an armband in the centre, through which the left arm was passed up to the elbow and the handgrip at the rim (**1**). The rim itself was offset, which could rest on the shoulder to help with the weight, especially when at rest. Held across the chest, it covered the citizen from chin to knee. However, being clamped to the left arm, it only offered protection to his left-hand side, though it did protect the exposed right-hand side of the comrade to his immediate left.

As in all military history, technology responded to the conflicts of the day and dictated what forms future battle would take, and with this new style of spear-and-shield warfare the weapon *par excellence* of our wealthy citizen was the long thrusting spear (Greek *dóru*, Latin *hasta*).

Our citizen also packs a sword. The introduction of the phalanx undermined the previous prestige of this weapon. Besides, in the crush and squeeze of a phalanx, a shorter weapon was preferable as it could be more easily handled. It may have required special skills to handle an antennae-type sword, but with a slashing-type sword it was almost impossible to miss in the cut and thrust of the tightly packed phalanx. One type was the Greek *kopis* (**2**), a strong, curved one-edged blade designed for slashing with an overhand stroke, not thrusting. The cutting edge was on the inside, like a Gurkha kukri, while the broad back of the blade curved forward in such a way to weight the weapon towards its tip, making it 'point-heavy'. Whatever the pattern, Greek or Italic, the sword was now very much a secondary arm – a far cry from its former predominance in the epoch of clan warfare – to be used only when a warrior's spear has failed him. It is worn suspended from a long baldric from right shoulder to left hip, the scabbard being fashioned of wood covered with leather, with the tip strengthened by a small metal cap, a chape, usually moulded to the scabbard.

1
2

Servius Tullius was remembered for his lowly birth (the son of a household slave) and a special relationship with the gods. Roman tradition has the king introduce the constitutional innovation of the census, which divided citizens into property classes. A clerk on the altar of Domitius Ahenobarbus (Paris, musée du Louvre, inv. Ma 975) records names, either as part of a census or as part of the levying of citizens for military service. (Fields-Carré Collection)

the property ladder. Thus class I contained 80 *centuriae*; classes II, III and IV 20 each; and class V 30. Below them were five *centuriae* of unarmed men, four of artisans and one of *proletarii* (citizens of the lowest class), whose property was too little to justify enrolment in class V. Known as *capite censi*, 'counted by heads', these men were simply counted and had no military obligations, no political rights and were not taxed. In other words poverty, curiously perhaps to us moderns, freed men from conscription. At the other extreme were those who served on horseback, the sons of the wealthy, making up 18 *centuriae*, which took precedence over the *centuriae* of the other five classes.

Servius Tullius was the first perhaps among Rome's rulers who realized that the days of individual combat were coming to an end. In doing so he had conceived of a state organized exclusively for war. Hence under the Republic the Servian system would provide the basis of the *comitia centuriata*, the 'assembly in centuries', at which all adult male Roman citizens with the right to vote did so to declare war or accept peace. They also elected the consuls, praetors, censors and senior magistrates (i.e. posts with *imperium*) of Rome, and tried capital cases. Gathering on the Campus Martius (Field of Mars), a sizeable open area located on the northern fringe of the city and enveloped by a bend of the Tiber, its structure exemplified the ideal of a militia in battle array, with men voting and fighting together in the same units. This assembly operated on a 'timocratic principle', the common idea whereby the property-owning classes lived in a 'stakeholder' society, where political rights are defined by military obligations, which in turn spring from the need to defend property; property itself gives the financial means to engage in that defence. Those who have property, and thus a stake and a role in the defence of society, are considered more likely to take sensible decisions about how the state is run. The richer you are, the truer this becomes, and conversely, having nothing to lose will make you irresponsible. The timocratic principle meant that only those who could afford arms could vote, which meant the *comitia centuriata* was in effect an assembly of property-owners-cum-citizen-soldiers. Oddly enough, the Servian army of Livy and Dionysios does not appear in their respective battle accounts.

The wealthiest citizen-soldiers of early Rome stood in the foremost ranks of the phalanx and wielded the *hasta*. This 5th-century BC Etruscan bronze spearhead (Arezzo, Museo Archeologico Nazionale) once formed the business end of a *hasta*. It has a leaf-shaped blade with a midrib and ends in a closed socket. The midrib gives greater longitudinal strength to a spearhead, increasing its effectiveness at piercing shields and armour. (Fields-Carré Collection)

EQUIPMENT AND APPEARANCE

Unfortunately for military historians, the Livian battlefield is a confused and contradictory place. On the other hand, excavations over the last century or so have produced a wealth of archaeological evidence, which enables us to build up a tentative picture of the early Roman warrior. Italic armies were fundamentally infantry armies, and pictorial evidence and archaeological finds make it clear that during our period of study, the principal weaponry available consisted of the sword and the spear. Other weapons included the javelin, the axe and the dagger. Warriors also had access, to varying degrees, to defensive equipment, notably helmets and shields. During the 9th century BC the objects deposited in graves were noticeably uniform, suggesting communities of people of relatively equal status. However, during the 8th century BC, especially in the later decades, the first signs of social differentiation are distinguishable. Grave goods became increasingly varied both in type and tone. There were more, and better, helmets, body armour and weapons, and the first imports began to appear, from southern Italy, Greece, Phoenicia and central Europe.

The shield was the most commonly used item of defensive equipment, but it was not the only one: helmets, pectorals and greaves were used to protect the head, torso and shins respectively. Most of what we know about these objects comes from examples that were fashioned in sheet bronze and have therefore survived, but it is highly likely that the majority of such items were made of more perishable materials, probably of wood, hide and bone. After all, these were materials that could have been used in a variety of ways, to give food and shelter, to adorn or to destroy. In addition, archaeology can tell us much about arms and armour, but never about the actual use of the weapons.

For a good illustration of this let us look briefly at the Capestrano Warrior (see photograph on p. 21). The warrior carries an elaborately decorated sword, and an almost identical one, some 60cm in length, was found in one of the 6th–5th century BC warrior

graves at Alfedena (130km east of Rome). A small knife is visible attached to the front of the sword scabbard. Similar knives, with a blade length of 20–25cm, have been found lying on top of sword scabbards in the graves at Campovalano di Compli near Teramo on the eastern flank of the Apennines. Similarly, on his upper left arm he wears an armband. Such armbands have been found in positions around the left humerus bone in the graves at Alfedena. This brings to mind one version of the famous tale of Tarpeia, the

OPPOSITE
The splendid finds (Rome, MNR Terme di Diocleziano, inv. 115194-207) from Lanuvium (Tomb of the Warrior) near Rome, dated to around 480 BC. The war gear includes a bronze muscle cuirass (with traces of leather and linen), a bronze Negau helmet (with eyes in glass paste, silver and gold) and a *kopis*. The athletic equipment includes a bronze discus, two iron strigils and a bottle for olive oil, a nice reminder of the dual aristocratic pursuits of war and sport. (URSUS)

LEFT
Bronze Villanovan helmet (Paris, musée du Louvre), second half of the 8th century BC. This pattern consisted of a bowl made from two halves, the join being decorated with a tall, arrow-shaped plate standing up from it. This solid crest had no practical use, though it did make the wearer appear taller and thus more frightening to the foe. It must have made the helmet somewhat top-heavy to wear too. The three long spikes projecting from reinforcing plates were purely decorative too. (Ancient Art & Architecture)

Roman maiden at the time of Romulus, who apparently betrayed the Capitol in exchange for the gold bracelets that the Sabine raiders wore on their left arms. She was instead crushed to death when the Sabines threw their shields – also carried on the left arm – upon her (Livy 1.11.6–9, Dionysios of Halikarnassos 2.38.3, Plutarch *Romulus* 17.2, Propertius 4.4).

For limited protection, the warrior wears circular breastplates and backplates joined by a broad hinged band, which passes over the right shoulder, and is secured into place by means of a harness in the form of leather cross-straps. Actual examples of these have been found at Alfedena in position on the skeleton in the grave. Some 20–24cm in diameter, they are made of bronze backed with iron. The hinges and other attachments are also made of iron (Connolly 1998: 101–02). Without doubt, Roman warriors wore a similar form of body armour (examples have been found as far apart as Ancona and Aleria), though the pectorals found in the Esquiline tombs are rectangular in shape with incurving sides, slightly less than 20cm wide and

Early Italic armour consisted of circular breastplates and backplates, some 20–24cm in diameter, and was the basis of Oscan armour. Affording a greater degree of protection than its rudimentary predecessor, the triple-disc cuirass consisted of three symmetrical bronze discs placed on the chest and back. This is an example from Paestum (Gaudo Tomb 174, *c.*390–*c.*380 BC) formerly belonging to a Lucanian warrior. His Attic helmet is on display too, complete with crest and feather holders. When an Italic warrior wore *aigrettes* in his helmet it was apparently in the hope of identifying himself with Mars (Virgil *Aeneid* 6.779, Valerius Maximus 1.8.6). (Fields-Carré Collection)

slightly more than 20cm long. The shorter sides are pierced with holes for the stitching of a leather backing and the attachment of leather straps to hold the pectoral in place. It protected only the upper chest.

Spear

Whereas the lordly sword was a weapon of military and political elites almost everywhere in the ancient world, the sine qua non of the physical brutality that we call heroic combat, the humble spear was much more of a common workaday weapon in almost all areas. This was due to its extreme simplicity of design and to its relative cheapness, using, as it did, a minimum amount of expensive bronze or iron in its construction.

Few ancient civilizations eschewed the spear in their arsenals, and in its simplest form, a spear was nothing more than a stout wooden stick with a sharpened and hardened business end, the latter being achieved by revolving the tip in a gentle flame till it was charred to hardness. Beyond that, it was a long, straight wooden shaft tipped with a metallic spearhead, although a sharpened stick with the hardened tip could be quite an effective killing weapon in itself. Ash wood (as frequently mentioned in the heroic verses of Homer and Tyrtaios) was the most frequently chosen because it naturally grows straight and cleaves easily. Moreover, it is both tough and elastic, which means it has the capacity to absorb repeated shocks without communicating them to the handler's hand and can withstand a good hard knock without splintering. These properties combined made it a good choice for crafting a spear.

It goes without saying that the inherent weakness of its wooden shaft was the main drawback of the spear in battle, and so to sustain some lateral damage in repeated use, a shaft had to be at least 22mm in diameter. Once the spear was made, it could provide long service with a minimum of maintenance. Additionally, it could be used in hunting as well as in warfare, thus providing a dual-purpose, versatile tool for our clansman.

As the main weapon of most clansmen, high and low alike, the spear, unlike the javelin, was usually employed as a far-reaching stabbing weapon, and as such it would rarely have left its owner's hand on the field of battle – a spear, once hurled, leaves the clansman defenceless. The simple term 'spearhead', however, embraces a great range of shapes and sizes, complete

These 4th-century BC ivory plaques (Rome, Museo Nazionale di Villa Giulia, inv. 13236-7) from Palestrina give a good impression of the war gear of an early Roman citizen. Much of it is Graeco-Etruscan inspired, and each man, armed with a pair of spears, appears to have a *clipeus* resting against their legs. These men would not have been out of place in the foremost rank of a Servian phalanx. (Fields-Carré Collection)

with socket ferrules (either welded in a complete circle or split-sided) to enable them to be mounted on the shaft and often secured with one or two rivets and/or binding. For use as a stabbing weapon, practical experience tells us that the width of the blade was important, for a wide blade actually prevents the spearhead from being inserted into the body of an enemy too far, thus enabling the spear to be recovered quickly, ready for further use. In our period of study the most common designs were angular blades with a diamond cross section, and leaf-shaped blades with a biconvex cross section.

Formidable bronze Villanovan spearhead, found at Cumae and dated from the late 7th century BC. One of the main problems of such massive spearheads was an increased weakening at the juncture between the blade and the socket. Though iron gave a keener cutting edge, spearheads of 'pitiless bronze' continued to be made for a very long time, bronze being easier to work and therefore cheaper as a finished product. (© Board of Trustees of the Armouries (VII. 1638 – DI 2005-0159))

The addition of a midrib gave greater longitudinal strength to a spearhead, increasing its effectiveness at piercing shields and armour during hand-to-hand spear play.

It is difficult to say with any certainty what the best length for a spear was, but common sense dictates that it would have been mainly between 2–3m in length. Any shorter and the chief advantage of keeping the enemy a whole pace away is gone; any longer and it becomes awkward and too wobbly to use accurately with one hand. Of course, it also had to be light enough to wield one-handed (usually overarm in a thrusting manner) and used in conjunction with a shield.

Though iron gave a keener cutting edge, spearheads of bronze continued to be made and used during the age of iron. Unlike bronze, iron was worked by forging rather than casting. All bronze weapons were cast from molten metal. Contrastingly, iron weapons had to be beaten into shape as it was impossible to obtain a sufficiently high temperature for the casting of iron. Beaten weapons were far stronger than cast ones, but the early technology of iron did not produce a commodity superior to bronze until a form of carburization was tried, leading to the production of a steel-like metal. It was the cheapness and comparative abundance of iron ores that was at the root of its popularity over bronze, which, after all, was a strenuous merger of copper and tin.

Sword

The sword originated from the realization that an extended dagger provided greater reach, which was more advantageous in combat. Of course, the superiority of the new weapon must have been demonstrated in grievous action before such a lesson was learnt. So the dagger slowly developed into a longer one, the dirk, and eventually the dirk was lengthened into the sword, which at first was relatively short.

CITIZEN-SOLDIER, CLASS III

As each citizen was obliged to buy his own equipment, it seems clearly logical to assume that not all hoplites were identically equipped. The less well-off citizen would have had nothing so elaborate as the bronze or linen corselet that wealthier citizens wore, yet doubtless many of those members of this class who had the means actually supplied themselves with a small bronze breastplate, the old Italic round or rectangular models being still very much in circulation (**1**).

The importance of armour to those who fight at close quarters can hardly be overstated. Apart from the obvious protection it offers, armour lends confidence to the wearer, and confidence in combat is always extremely important. Where metallic armour was not available or affordable, citizens probably made use of *cuir bouilli* or padded protection, and we can be certain that the individual citizen-soldier sought to protect himself with at least some form of body armour. The private provision of (expensive) war gear could accordingly reflect individual preference for different forms and styles (Greek, Graeco-Etruscan or Italic).

For the most part, however, to compensate for any lack of body armour classes II and III used the oval *scutum* instead of the round *clipeus*. The *scutum* offered better protection to the torso and legs, it being the body shield common in Italy and already known in Rome as it had been widely employed in its early days by its clan warriors. In shape and form, whatever may have been true of the period of the clan-based warband, the *scutum* would by this period have been very much like the *thureós* ('door-like') common to the soldiers called *thureophóroi* in later Hellenistic armies. With the *scutum* a soldier could be both defensive and offensive, parrying enemy blows with its board or rim and punching with its metallic boss-plate. The *scutum*, unlike the *clipeus*, was a relatively cheap piece of equipment.

1

Leaf-shaped double-edged bronze Villanovan sword and scabbard, found at Cumae and dated from the late 7th century BC. Swords were habitually buried in their scabbards. Sometimes these were further wrapped in linen cloths. Their purpose was to provide maximum protection for the valuable weapons. Conveying as it did considerable status and symbolic significance, the sword served as both an item of display and a weapon of violence. (© Board of Trustees of the Armouries (IX. 1280)

Bronze antennae sword (Bologna, Museo Civico Archeologico, inv. 77001) from the Sepolcreto San Vitale Tomb 776, dated from the end of the 8th century BC. Here we have a detailed view of the hilt, showing the part normally known as the pommel coiled over to form the diagnostic volutes that give this sword pattern its name. The grip and guard are likewise of cast bronze. In Bologna, as elsewhere, swords were buried only in the highest-ranking male tombs. (Fields-Carré Collection)

As the sword grew in length it also became a symbol of power and lordliness. At the same time, however, it was also the high-status warrior's most important weapon. A high-status warrior with a long sword was equipped for close-quarter fighting, for heroic man-to-man battles in which the opponent could be cut down at close quarters. Not surprisingly, the sword was his proudest possession. In evaluating the calibre of a sword, two factors were of crucial importance: the correct positioning of the centre of gravity, and a well-executed juncture between the tang, the extension of the blade over which the parts of the hilt (the guard, grip and pommel) were slotted, and the blade itself. A centre of gravity nearer to the hilt made for a more efficient weapon for both cutting and thrusting, while the juncture between tang and blade was an area of potential weakness. A warrior needed to be confident that his blade would not bend from the tang, nor his grip loosen, no matter how jarring a blow he struck. To fashion the perfect sword was a true challenge for the sword smith.

In the age of clans two sword types can be distinguished: the short, broad-bladed sword, which was very robust and well suited for slashing, and the longer, narrower-bladed sword, which could be used to thrust as well as to slash. Both types tended to be of bronze, while the latter generally had a tang with 'antennae' as opposed to a pommel as such, which sometimes coiled over to form distinctive volutes. The grip was fashioned either from organic matter, in which case it was not preserved, or from metal. Its pieces were seated in place within the guard (usually a metallic plate) and the pommel and attached through the tang by metal rivets.

A thoroughly efficient cut-and-thrust weapon, the sword known to specialists as the 'antennae type' originated in central Europe and is closely related to the Naue Type II sword. It was a long weapon (most surviving examples are some 70cm from pommel to tip), the majority having a leaf-shaped blade that ends in a sharp point. Blades that were leaf-shaped in form naturally added more weight to a downward cutting action, so their effect was often more devastating. The antennae-type sword could be used as a thrusting weapon too, but obviously it was designed primarily for slashing. In swords whose primary purpose was for thrusting, the centre of gravity was just beyond the hilt. On the antennae-type sword the centre of gravity was much farther along the blade, which thus added greatly to the force and velocity of a slashing blow.

Shield

Clan warriors attempted to protect themselves with a body shield, the Italic *scutum*, carried in the left hand. This afforded some shelter from missiles, spear thrusts and sword slashes. It was their primary defensive aid, as very few men at the time could afford to equip themselves with helmet or body armour. In skilled hands, a wood-and-leather shield would have deflected a sword blow or absorbed the impact from a spear, but it did not offer complete defence. Agility was the surest weapon of the lowly clansman.

Cut through the centre of the shield board was a circular hole. Covering this hole was the thickest part of a long central wooden spine strengthened with a sheet-metal boss-plate riveted to the shield board. This afforded security to the left hand on the inner side, which was gripping a horizontal handgrip. In the form of a wooden crossbar, this handgrip was plausibly bound with cloth or leather.

Archaeologically, shields are uncommon, being as they were mainly of a wood-and-leather construction. However, one shield of the *scutum* pattern was discovered at Kasr-el-Harit, a small town in Egypt, remarkably well preserved in the dry sands of Fayûm (Connolly 1998: 132). It is midway between a rectangle and an oval in shape, and is 1.28m in length and 63.5cm

We have a marvellous depiction of hoplites going into battle on the Protocorinthian olpe (Rome, Museo Nazionale Etrusco di Villa Giulia, inv. 22679), found (1881) in an Etruscan tomb on the estate of Prince Mario Chigi near Cesena. It is dated to around 650–40 BC. This shows the basic characteristics of the new shield, the *aspis*, which in no way resembles earlier forms. It was this large double-grip shield that covered the hoplite from chin to knee, and more than anything else made the phalanx possible. (Fields-Carré Collection)

Arnoaldi *situla* (Arnoaldi Necropolis, Grave 2), a bucket-shaped container of sheet bronze, dated to around 450 BC. Two zones of repoussé depict a military parade of chariots and warriors, the latter carrying the Italic body shield, or *scutum*. The *scutum* had only a single, horizontal handgrip in the centre, protected by a large metal boss-plate. This allowed it to be moved about freely, and the boss-plate could be used offensively too, by punching the enemy. (Ancient Art & Architecture)

in width with a slight concavity. It is constructed from three superimposed layers of birch laths, each layer laid at right angles to the next. The layers were glued together, and originally the finished plyboard was covered with lamb's-wool felt. This was likely fitted damp in a single piece, which, when dry, had then shrunk and strengthened the whole ensemble. The shield board is thicker in the centre and flexible at the edges, making it very resilient to blows, and the top and bottom edges may have been reinforced with bronze or iron edging to prevent splitting. Nailed to the exterior face and running vertically from top to bottom is a solid wooden spine in three sections. Except for the nails, no other metal components, such as a boss-plate, were found. Its plywood construction gave a far greater degree of battleground resilience than its plank equivalent. Nevertheless, thickness was compromised in order to give the bearer a large manoeuvrable shield.

Citizen phalanx

The Servian system certainly suggests the existence of a settled agrarian society with its citizen militia based upon the Greek hoplite phalanx, for the links between citizenship, non-movable property and military obligation were fundamental ingredients in such a socio-political system. Now the army of Rome was simply the civil population under arms, the civilian and the soldier being the same man in two different aspects. It was the duty of the citizen to also be a soldier; soldiering was a branch of citizenship and warfare was a branch of politics.

Political science aside, at the tactical level this change in the pattern of Roman warfare resulted in a shift towards spear-and-shield combat (with rather cheap weapons) and the tactics connected with Greek-style phalanx fighting in the open field. However it came about, the phalanx, with its spearmen similarly equipped and fighting shoulder to shoulder, was concomitant with the rise of poorer but focused and highly competitive societies, city states in which the hoplite, with his very costly equipment, was a citizen of some property. Yet Rome, in order to increase the manpower resources for its new-model army, went a step further by allowing those of smaller means to become full members of its society. Though the 'good and substantial' citizens had a greater share in expressing and formulating the 'will of the people' than the 'unwashed', the latter too belonged to the polity called Rome.

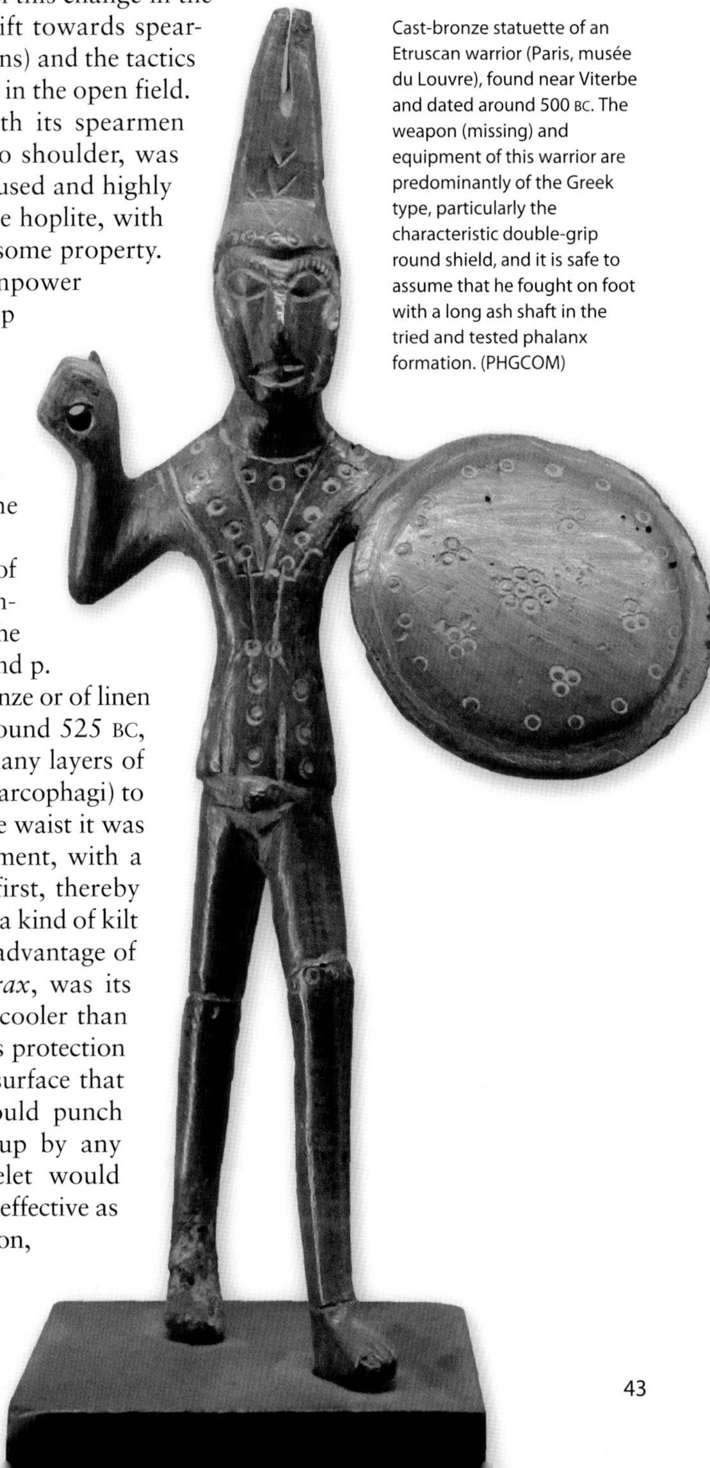

Cast-bronze statuette of an Etruscan warrior (Paris, musée du Louvre), found near Viterbe and dated around 500 BC. The weapon (missing) and equipment of this warrior are predominantly of the Greek type, particularly the characteristic double-grip round shield, and it is safe to assume that he fought on foot with a long ash shaft in the tried and tested phalanx formation. (PHGCOM)

We will now deal quickly with the subject of arms and equipment for our new citizen-soldier, the meat of this being dealt with in the commentaries of plates C and D (see p. 30 and p. 38). For the citizen of means, a corselet of bronze or of linen fully protected his torso. First appearing around 525 BC, the latter type of corselet was made up of many layers of linen glued together with resin (cf. Egyptian sarcophagi) to form a stiff shirt about 5mm thick. Below the waist it was cut into strips, *pteruges*, for ease of movement, with a second layer of *pteruges* fixed behind the first, thereby covering the gaps between them and forming a kind of kilt that protected the wearer's groin. The great advantage of the Greek-style linen corselet, the *linothôrax*, was its comfort, as it was more flexible and much cooler than bronze under a Mediterranean sun. As far as protection goes, the main advantage of bronze was a surface that deflected glancing blows. A direct hit would punch through the metal, but it might be held up by any padding worn underneath. A linen corselet would not deflect glancing blows, but it would be as effective as bronze against any major thrust. The protection, then, was slightly less than that of bronze, but the advantages of comfort and weight overrode that consideration.

Tufa cinerary urn (Florence, Museo Archeologico Nazionale, inv. 5744) from Volterra, 2nd century BC. The relief depicts two Etruscan warriors, one bearing a *clipeus* (right) and the other a *scutum* (left). Both these shield types were used in the Greek-style phalanx of early Rome, the *clipeus* by citizen-soldiers of class I and the *scutum* by citizen-soldiers of classes II, III and IIII. (Fields-Carré Collection).

BELIEF AND BELONGING

It was nervous awe of the gods, so says Polybios (6.56.7), that promoted the cohesion of the Roman state. True or not, from their earliest days the Romans, like so many other agrarian peoples, were both superstitious and religious, believing in the existence of numerous deities, each of whom possessed specific powers exercised over discrete aspects of the physical world. Unlike the Greeks, however, the Romans did not isolate their gods in the privacy of their individual temples. Nor did they develop a complex and colourful mythology, with its cosmogony, celestial marriages and incestuous

Italic Negau helmet and Graeco-Etruscan greave from Brisighella, Ravenna (San Martino Tomb 10), 5th century BC. Albeit Umbrian in context, such equipment would not look out of place in the Servian phalanx of early Rome. The use of Italic armour, in this case the helmet, hardly affected the function of the Greek-style phalanx as long as the front rank citizen-soldiers bore the *clipeus* and *hasta*. (Fields-Carré Collection)

genealogies, they simply conceived of their gods in rather practical terms as powerful entities, whom they diligently revered in order to receive benefactions and to avert bane. These obviously included the Capitoline triad of Iuppiter Optimus Maximus, Iuno Regina and Minerva, and, of course, Mars.

Gods of crops and war

Hand in hand with the idealistic portrayal of war and warriors is grim realism, with the result that ritual and superstition become interlocked with the pursuit of war. Thus most warring societies developed gods specifically devoted to war. Mars was a very important god among early Romans as well as for the other Italic peoples of early Italy. In later times he was regarded as the principal god of war, being a far more Olympian figure than his Greek counterpart Ares, but his nature was much more complex in archaic times. He may have been the god of the wilderness lying just beyond the fences of the peasant's home, who therefore was thought to exercise power over both field and forest, cultivated land and untamed nature. The symbolic spaces of the hearth, homestead or hamlet were starkly contrasted with the dangerous territories outside, of forest, mountain and marsh. Moreover, the peasant's life was hard and insecure; indeed, his very existence was precarious. Bad weather, crop failures or his being away at war (the campaign season was high summer) was always uppermost in men's minds. Consequently, Mars was invoked to protect Roman crops and to assist Roman forces in waging war beyond the margins of their home.

Mars' dual agricultural and warlike nature is indicated by the fact that the early Romans began the year with the month of March, *mensis Martius*, which took its name from the god and marked the return of spring, plant life and the campaigning season. The Salii, the leaping priests, performed their dance through the dusty thoroughfares of palisaded Rome during this month, beating spears upon shields and chanting an archaic hymn, the *Saliare Numae carmen* (Horace *Epistula ad Pisones* 2.1.86), which was in fact unintelligible in the poet's day. They ended the sacred day with a sumptuous feast (Horace *Odes* 1.37.2). Livy, who attributes their introduction to King Numa Pompilius, says they wore 'a bronze covering for their chest' (1.20.4), perhaps a rectangular pectoral, and Plutarch (*Numa Pompilius* 13.4) adds that they wore bronze studded belts and bronze helmets, but substitutes the spears for

Mid-6th-century hoplite panoply (Olympia, Museum of Archaeology), consisting of a bronze bell-shaped corselet (breastplate left, backplate right) and the bronze facing of an *aspis* (centre). This type of corselet took its name from the flange, which flared outwards below the waist like the mouth of a bell. The flanging helped to deflect incoming blows. The hammered sheet bronze for this type of corselet was generally 0.6–1mm thick, giving a weight of about 5kg. The breastplate overlapped the backplate, being secured by external hinge pins. (Fields-Carré Collection)

Scene from the Amazonmachy decorating a marble sarcophagus (Florence, Museo Archeologico Nazionale) from Tarquinia, mid-4th century BC. Here we see a splendid depiction of the *linothôrax*, a linen corselet cut in the Greek style with *pteruges* and tied-down shoulder doubling. The Amazons are wearing the 'Thracian' helmet. (Ancient Art & Architecture)

short daggers. After the hard months of winter, the growth of crops was supposed to be encouraged through the sympathetic magic of their leaping, and their hymn commemorated the passing of the old year's spirit of growth (Veturius Mamurius: Old Mars) and the return of the new year.

E

RAID, SABINE SETTLEMENT, *c.*750 BC

Inter-community conflicts in our period of study were more like vicious raids than pitched battles, and often resulted in women and children being captured, livestock being stolen and the elderly being murdered. The task of guarding the social group against banditry and plundering, seizing goods from neighbouring settlements and even challenging others to fight would have become the responsibility of specially selected individuals, skilled in handling weapons and organizing others.

The clan was formed out of a larger mixed following centred around a small elite. The latter would normally consist of a ruling family, whose status might be hereditary, but in practice depended on an aura of success, in military matters of course but also by holding a commanding economic position within society. The mutual interdependence of the components of the ruling elite required the supreme king to be seen to be generous in his gifts of lands, slaves, booty and other resources to the other clan rulers, who in turn redistributed some or much of what they received to secure the continued allegiance of their own followers. Successful warfare, too, could play a crucial part, in that it provided opportunities for the members of the great families, those that later became known as patricians, to emulate the deeds of their heroic forebears, real or imaginary. Naturally, the choice of which of the clan chieftains should rule Rome (as 'king') was primarily in the hands of these men.

Fighting and looting were closely allied, and both were considered fit occupations for the best of men. Production was work for peasants or slaves: men of the warrior caste helped themselves to the fruits of others' labours. The numerous rapacious raids that form the background to the early history of Rome were in truth a protracted series of robberies with menace. A settlement might be raided, land ravaged, crops and buildings ruined and storehouses ransacked. Here we witness one such attack on a Sabine hilltop hamlet. Its men have fallen defending hearth and home, and the Roman raiders are now consigning the settlement to a bloody plunder, killing its elderly, abducting and enslaving its womenfolk and dragging off all its portable wealth. Where there had once been a busy Sabine commune, the Romans will leave behind a ruinous heap.

Group identity

Every people has certain traditionary and religious ideas, sustained by marvellous myths and stirring stories, which underlie its institutions. Despite our modern disregard of tradition, and even amid the innumerable influences affecting modern civilization, each of our nations moves within a charmed circle of its own. But when religious, social and political ideas are inextricably woven, springing from one common root, as in the case of Rome, it may be believed that the influences they exercise become sacred – what the Romans respected as the *mos maiorum*, their ancestral traditions – and something quite beyond the experience of our laic nations.

Let us look at, for instance, 5 July in the Roman calendar, the public festival known as the *Populifugia*, 'the day of the people's flight'. This was a primitive military ritual involving the assembly and purification of Rome's adult male population (probably under arms), followed by a ceremonial rout of Rome's foreign enemies. The origin of *Populifugia* is not known. The very learned antiquarian Varro (*Lingua Latinae* 6.18) thinks that the ceremony commemorated the flight of the dispirited Romans shortly after the sack of Rome by the Gauls.

The *Populifugia* would be later eclipsed (and driven into obscurity) by the censorial lustrum at the end of a census and by the *Transvectio Equitum*, Procession of the Horsemen, of 15 July, instituted by the censors of 304 BC (Livy 9.46.15). The best description of this military festival is given by Dionysios of Halikarnassos (6.13.4), who states that the parade started at the temple of Mars located about a couple of kilometres beyond the Porta Capena on the Via Appia, in a grove where the army assembled before marching off to war. On entering the city, the parade made its way to the temple of Castor and Pollux in the Forum, an obvious choice, and then on to the Capitol.

The end of the campaigning season was marked by a ceremony in honour of Mars, the *Armilustrium* of 19 October, in which men assembled fully equipped under their standards and underwent ritual purification. The war gear and standards were then stored for the winter months. It appears that on this occasion captured arms were dedicated to Mars too.

This 4th-century BC bronze statuette (Paris, musée du Louvre, Br 124) of a Samnite warrior is believed to have been found in Sicily, and thus possibly represents a mercenary serving there. Alternatively, it may well represent the war god Mamers (Mars). Either way, he wears an Attic-type helmet with a groove and sockets that once held crest and *aigrettes*, characteristic Oscan triple-disc cuirass, broad Oscan belt and Graeco-Etruscan greaves. The shield and spear he once carried are lost. (Ancient Art & Architecture)

ON CAMPAIGN

Latium was a raw, new land, a pioneer country where danger and opportunity were equally abundant. It was a land of wild beasts, outlaws, cattle raids and blood feuds, a land split into many civic communities more or less like the palisaded village of Rome. Among them, between each community and every other one, the normal relationship was one of hostility, at times passive, in a kind of armed truce, and at times active and bellicose. Even within a community there could be anarchic competition between rival families and clans. Hard times make hard men, and the warriors of Romulus' world were plundering, warlike men whose swords did not hang idle from their belts. Such was the warfare practised by peoples living below the institutional level of the state.

While it is not always sensible, when faced with small and sundry morsels of evidence that have survived largely thanks to chance, to assume that they must be somehow linked and capable of being united to tell a single and coherent tale, in the case of the sources relating to the murky opening years of Rome there may be some merit in such an approach. No doubt, minor skirmishes were remembered as major engagements, and grew bigger and bigger in the recounting of them. All the same, skirmish or otherwise, for our clan warrior it was a very real and a very dangerous place, a place where he confronted cold death and its remorseless appetite.

For military purposes of a colonial nature, on the other hand, it was considered appropriate to take a census of the citizens and to reclassify them on the basis of wealth and age. And so with the Servian reforms the Roman army, now several times greater in size and modernized in its mode of fighting, became a 'nation in arms'. Yet citizen-soldiers were, by and large, still farmers and could afford to spend only a few summer weeks on the campaign trail before they needed to return to their fields. As a result conflicts were of a short duration. If the new weaponry and tactics of the phalanx meant war was a much more bloody and pitiless business than the raid and ambuscade of earlier days, it also allowed for a high measure of decision in battle, certainly a prerequisite for outnumbered but well-organized citizen-soldiers. It also allowed for successful territorial expansion, as in the case of Rome.

Italic votive bronze plaque (Atestino, Museo Nazionale), 5th century BC, depicting a warrior bearing a *clipeus*. The rest of his hoplite panoply consists of an Italic pot helmet with fore-and-aft horsehair crest, Graeco-Etruscan greaves and two spears, one apparently with a head larger than the other. (Fields-Carré Collection)

Detail on 3rd-century BC alabaster cinerary urn (Palermo, Museo Archeologico, inv. 8461) depicting two Etruscan warriors wearing Etrusco-Corinthian helmets. The Etrusco-Corinthian pattern was clearly derived from the Greek Corinthian helmet but worn like Perikles on top of the head much like a pot helmet. Thus, the eyeholes and the nasal guard were purely decorative. It could have cheek pieces. A crest could be raised on a stilt, and feather holders added at the sides. (Fields-Carré Collection)

Raid and ambuscade

Rome's early wars were then little more than sudden smash-and-grab raids rather than the wholesale slaughter of an enemy, and its first warriors – proud, rude and barbarous – belonged to armies that were little more than brigand bands led by 'robber barons'.

Nietzsche once advocated living dangerously, extorting his readers to be 'robbers and ravagers as long as you cannot be rulers and possessors', to seek out conflict in order to experience grandeur (*Die Fröhliche Wissenschaft* §283). It is as if this philosopher of the will to power was acutely aware of the mentality of illiterate foemen-freebooters such as the semi-divine Romulus, and the equally illiterate cattle-rustling followers. These men with near subsistence-level lifestyles certainly wanted to live dangerously, hunt and plunder; they were warriors, and the main theme of a warrior culture was constructed around two concepts: prowess and honour. The one is the warrior's essential attribute, the other is his essential aim. These men wanted to get rich: they were looters and pillagers, unprovoked raiders of peaceful settlements during the brief fighting season. They were, judged by modern standards, brazen-faced bandits greedy for loot. Not for the first time, war was brigandage carried on by other means.

The fighting season was between the grain harvest and the sowing. Italian agriculture involved seasonal hard work – ploughing, sowing and milking herds – a hurried harvest and long periods of relative leisure. Under the sign of Mars, the Romans attacked a settlement, stealing, preying on the powerless and burning. War in this period was not typified, and certainly not decided, by the outcome of battles. They left behind them a lamentable spectacle of scattered and smoking ruins, deserted, sulphurous fields and empty thoroughfares. They stripped their target until it was bare of all moveable goods, a baleful wasteland of stony rubbish.

Italic Negau helmet (Arezzo, Museo Archeologico Nazionale), 5th century BC. The Negau type was a bronze bowl with a rib running fore and aft and a lateral depression at its base. It had no cheek-pieces and was held in place by a leather chin strap. Developed in the 6th century BC from the Italic pot helmet, the Negau pattern remained in use unchanged down to the 3rd century BC. (Fields-Carré Collection)

Pitched battle

A formal battle was probably an uncommon experience for the pre-phalanx Roman army, simply because that was not the nature of the game in this period. Nonetheless, when they did occur, a full-scale, set-piece battle was a horrendous thing. Getting the two armies to close and commence this maelstrom may not have been easy. Howsoever the attack developed, those in the foremost rank would have lowered their spears, hunkered down behind their shields, and prayed to their gods.

Wars are the sum of battles, and battles the tally of individual human beings killing and dying, and though the individual comforts himself with the belief that death might come to the next man, and not to him, concrete realities ultimately decide whether soldiers return home in safety or are left on the battlefield to be spoil and booty for the carrion birds. For one thing, Mars often fights too haphazardly to give each one his due. For another, although a single war leader may be morally and legally responsible for starting a war, the actual fighting – the winning and losing – is done by hundreds and thousands of unknown soldiers bereft of identity. History is rarely generous in its attentions to men of small consequence, yet Tolstoy, in his magnificent *War and Peace*, teaches, among many other things, that battles are fought by Pierres and not by the Napoléons. Moreover, the losers in war are not merely disappointed – they are wounded, impoverished, enslaved, raped, orphaned, widowed or just plain killed.

A cutting, slashing or crushing blow does not need to penetrate armour in order to cause damage – contact is enough. For if the blow is violent enough, this force will be transferred through the armour causing blunt-trauma injuries, including broken bones and internal haemorrhaging. Mid-form Corinthian helmet (London, British Museum, inv. GR 1881.7-25.1), dated *c.*600 BC, showing obvious battle damage. (Fields-Carré collection)

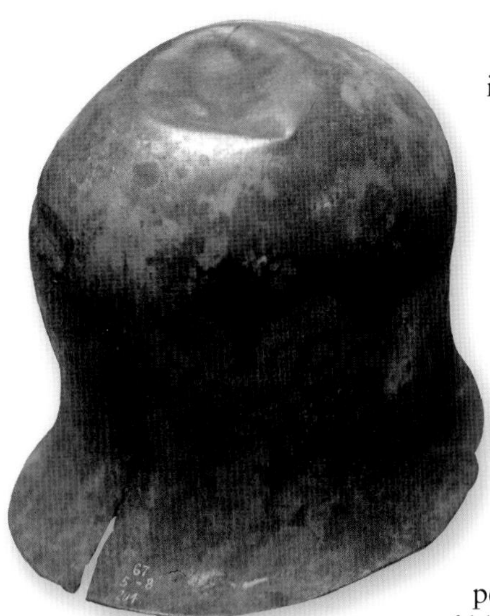

Meanwhile back on the field of battle itself, there is one undisputable, unwelcome fact that death, the great equalizer, might come visiting at any instant. Only the choice of whether to fight or flee confronted the Roman warrior now as an escape from the darkness of ultimate destruction. It may be assumed that at this point the sensible man, not wishing to waste his manhood, started running, for when that critical moment comes his legs will tremble and weaken, his heart will rush and pound, and he will turn and run. Rational cowardice, perhaps, but the language of war is that of blood and flesh.

F

VICTORY, LAKE REGILLUS, 499 BC

It goes without saying that an important part of the story of Rome is the long series of bitter wars by which it subdued the peoples of Italy. However, at first the Romans were preoccupied with the circle of peoples in their immediate neighbourhood: the Etruscans to the north, the Latins to the south and the Sabines, Aequi and Volsci to the east and south-east. It is the Latins that concern us here.

With the abolition of the monarchy and the collapse of Etruscan power south of the Tiber, Rome claimed to uphold the hegemony in Latium. The Latins, naturally, refused to tolerate this arrogance and quickly organized themselves into a league from which the Romans were excluded. The league forces, led by the erstwhile king Tarquinius Superbus and his son-in-law Octavius Mamilius of Tusculum, met the Romans at the battle of Lake Regillus, close to Tusculum. Although Livy, in his stirring narrative (2.19.3–20.13), livens up the ensuing fray with hints of the divine presence of those inseparable heroes Castor and Pollux, the Heavenly Twins, the engagement itself remains a historical fact, though it was hardly a glorious victory for the young Republic.

The story of the Heavenly Twins seems infantile and we wonder why Livy bothered to record it. Yet the account given by the Augustan historian is one that has been reshaped by tradition, myth mingling with reality, and the resulting peace accord, the *foedus Cassianum*, saw Rome, which had nothing or at most little to brag about, formally resign any claim to hegemony in Latium and recognize its position as an equal of the Latin communities (Dionysios of Halikarnassos 6.95.2, cf. Livy 2.33.4). It also heralded a comparatively peaceful period, which lasted for the greater part of the 5th century BC. And as for Tarquinius Superbus, he packed his regal bags and went off into exile at the court of Aristodemos, the tyrant of Cumae.

We are reliant on literary sources for our information, so battles fought by tribal peoples, the early Romans included, who did not commemorate them go unrecorded. Hence Lake Regillus became a 'battle worthy of note', and though his account is written in epic and poetic terms (as Lord Macaulay noted), Livy (as did Dionysios of Halikarnassos) considered it to have been one between hoplite phalanxes, the Latin one being bolstered with Roman exiles. At some point in the engagement, when the day looked like turning bad, what Livy calls the *equites* dismounted and joined the exhausted Roman phalanx to eventually carry the day. As in most archaic Greek states, these were probably mounted hoplites as opposed to true cavalry – aristocrats who rode to war only to dismount and fight alongside their less-wealthy fellows. Given the association of the Heavenly Twins with horses and horsemen, the battle was proclaimed by the later Romans to have been won by the *equites*. Be that as it may, horses played a minimal role in Roman warfare, and Rome relied, as it was to do for the centuries to come, on its foot soldiers. The role of the *equites* as horsemen in warfare was restricted to impressive careering gallops along the battle lines rather than for cavalry charges.

Here we join the battle at the moment the *equites*, having jumped from their horses, join the depleted ranks of their weary comrades. The new arrivals are rushing into the fray, their burnished bronze corselets conspicuous among the blood-splattered ones all around.

Livy supplies next to nothing about the nature of command or tactics during the battles he describes. We hear of commanders giving their men pep talks, of armies colliding and clashing and of the occasional shout of encouragement. But a great deal is left to brute force, courage and chance. We shall assume that a clan chieftain was such because in time of war he fought with conspicuous courage, admirably demonstrating that heroic ideal to step out in front of all others and purposely mix it up at close quarters in the middle of the battleground with an enemy chieftain, challenging, fighting and winning within sight of friend and foe alike. We should not think of him as sitting back quietly on a hill – or on a horse – judiciously overseeing the action from a safe spot behind the battle line, monitoring developments and despatching terse messages to his subordinates. Competition, especially in battle, and particularly in its close-quarter climax, presented him with the opportunity to acquire or reinforce the prestige and legendary status he needed in order to strengthen and maintain his pre-eminent position within the highly stratified social structure of his own close-knit clan. In the words of Livy, 'It was a point of honour in those days for the leader to engage in single combat' (2.6.9). And so in an age when intellectual pursuits were poorly regarded, leadership tended to be physical, robust and violent.

Prataporci from Monteporzio, not far from the ruins of Tusculum in the Alban hills, the supposed site of the battle of Lake Regillus. The lake, the relic of a volcanic crater, was drained for farmland in the 17th century. (Magnus Manske)

The second form of combat is that of the general fight on the battlefield, into which common clansmen would be drawn. We are assuming that there were few niceties of tactics, instead picturing nothing more than a sprawling scuffle and scramble of men as each side painstakingly hammered away at each other until exhaustion or weight of numbers swung the balance. In this desperate adrenalin-fuelled scrimmage of hacking and thrusting, combat survival not only relied on moving the shield promptly to block sudden blows from unexpected directions, but also, to an important degree, depended on athletic ducking and weaving to avoid potentially fatal blows. Shields were expendable. Intended to deflect or absorb blows, they would often have been damaged or destroyed in battle. There were precious few laurels to be won for the lowly clansman when hundreds of his comrades lay dead or dying on the field.

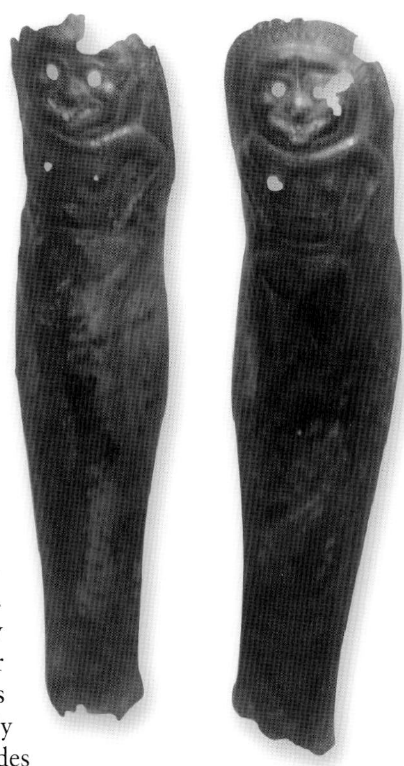

Pair of bronze greaves (London, British Museum, inv. GR 1856.12-26.615) from Ruvo, Apulia (*c.*550–*c.*500 BC). The hardest part of the body to protect with the shield were the lower legs, and, although not in itself fatal, a blow to the shins could prove debilitating enough to allow the citizen-soldier's guard to slip – thus exposing him to a killing blow. Though cumbersome to wear, greaves protected the shins and followed the musculature of the calf. On both knees, this particular pair feature the popular apotropaic motif of the gorgon head. (Fields-Carré Collection)

A detailed study of the male skeletal remains from the Alfedena necropolis has concluded that the nature of most of the cranial wounds (large ones from blades and small ones from projectiles) and their severity suggest that they were the result of combat. The trauma pattern for these unlucky individuals indicates that the blows to the head came from all directions, and it is suggested that these extremely violent injuries were probably inflicted by warriors from rival Samnite communities who had raided the victims' settlements (Paine *et al.* 2007).

Warriors are not soldiers. Both are killers, and both can be courageous, but disciplined soldiers value the group over the single heroic warrior. As such, they can operate en masse as a collective whole. Clan warfare, with its ancient allegiances of kinship, had given rise to confrontation and duels characterized by fervour and fury. For this reason, the advent of the Greek-style phalanx, with its armoured spearmen fighting shoulder to shoulder in a disciplined formation, changed the very nature of combat: individual exploits were replaced by corporate actions. And so in Rome, the archaic clan warrior became a disciplined citizen-soldier.

This is perhaps the time to take a look at the old chestnut of how the phalanx fought. It was the *clipeus* that made the rigid phalanx formation viable. Half the shield protruded beyond the left-hand side of the bearer. If the man on the left moved in close he was protected by the shield overlap, which thus guarded his uncovered side. Hence, hoplites stood shoulder to shoulder with their shields locked. Once this formation was broken, however, the advantage of the shield was lost. As the Greek Plutarch, writing a century and a half after Livy, says (*Moralia* 220A12), the body armour of a hoplite may be for the individual's protection, but his shield protected the whole phalanx.

The phalanx itself was a deep formation, normally composed (in the Greek world, at least) of hoplites stacked eight to 12 shields deep. A common hypothesis is that a Roman phalanx had its first two ranks made up of class I citizens, the third and fourth ranks of class II citizens, the fifth and sixth ranks of class III citizens and the seventh and eighth ranks of class IV citizens). In this dense mass only the front two ranks could use their spears in the mêlée, the men in the third rank and farther back adding weight to the attack by pushing to their front. This was probably achieved by shoving the man in front with your shield. Thucydides (4.43.3, 96.4) and Xenophon (*Hellenika* 4.3.19, 6.4.14), Greek authors who had first-hand experience of hoplite battles, frequently refer to the push and shove of a hoplite mêlée.

G

SURRENDER, THE CAUDINE FORKS, 321 BC

The Samnites were perfectly capable of mobilizing themselves and federating into a league when they needed to fight, and these rugged highlanders made formidable foes. The Romans knew something of this, as they had faced them in long, savage wars, sometimes accompanied by brutal reverses, as happened in the pass known as the Caudine Forks near Caudium in western Samnium, when the entire Roman army suffered the humiliation of being forced to pass under the yoke (Livy 9.6.1–2). A very Italic symbol of defeat, this was a frame made from two spears stuck in the ground with a third one lashed across horizontally at a height that compelled the Roman soldiers, who were disarmed and clad only in their tunics, to crouch down underneath. Having done so the last shreds of self-esteem and security were stripped from the individual. This ignominious disaster, which ranked alongside the Allia debacle, was to be the last time that Rome accepted peace as the clear loser in a conflict.

Inconclusive skirmishes and frontier raids were the order of the day for the first five years of the Second Samnite War. These speedy strikes were undoubtedly carried out by comparatively small bodies of men, with the consuls acting independently of each other. In order to end this five-year impasse it appears the Romans attempted to adopt a more pugnacious stance against the Samnites, pooling the forces of both consuls and advancing into the territory of the Caudini, the most westerly and therefore the most exposed of the Samnite tribes. Livy says (9.2.5) that the Romans were on their way to Apulia, but more likely this may have been an attempt to force a decisive success by knocking the Caudini out of the war. Whatever their intentions, they advanced into the Caudine Forks, two narrow wooded defiles with a grassy vale between them. The Samnite League generalissimo, obviously well informed about the Romans and their intentions, hid his men and blocked the further defile with a defended barricade of felled trees and boulders. The Samnites' successes in mountainous and difficult terrain confirm what Cicero implies (*De Oratore* 2.325), namely, that these doughty highlanders employed a flexible and open order of fighting, instead of relying upon a close-packed phalanx.

When the marching Romans reached the barricade they made an attempt to carry it, failed, and retraced their steps in haste, only to discover that the defile by which they had entered was also blocked with its own defended barricade. They were excellent fighters down on the flat lands or out in the open countryside, but less able in these unfamiliar mountains. After vain attempts to cut their way out, realism broke out as nervousness and then panic stepped in, and as fear began to lay its cold hand upon them the consuls surrendered, ostensibly to avoid starvation.

According, at any rate, to Livy, the Samnites had no idea what to do to take advantage of their spectacular success. Hence their leader, Caius Pontius, sent for his father, Herennius. When he arrived he explained that if they were to set the Romans free without harm then they could terminate the war there and then on equal terms. If they drove home their victory by killing every last one of them then Rome would be so weakened that it would cease to pose a threat for years to come. At this Caius Pontius asked was there not a middle way. The father insisted that any middle way would be utter folly and leave Rome smarting for revenge without weakening it. Ignoring his father's Nestor-like sapience, Pontius made the Romans suffer the yoke. Even if only a parable, Livy's account serves as a powerful illustration that the middle way is not always the best.

Bronze figurehead (Paris, Musée national de la Marine, inv. 41 OA 74) of the ironclad *Brennus*, launched in 1891. In 390 BC Rome was sacked by Brennus. When the Romans expressed displeasure to him that he was using fixed weights to enlarge the agreed ransom, the Gallic chieftain flung his sword into the weighing scale with the stern words, 'Vae victis!', or 'Woe to the vanquished!' (Livy 5.48.9). True or not, a more apt riposte cannot be imagined. (MED)

In hoplite warfare, therefore, the phalanx was the tactic. When one phalanx squared up to face another, and here we are assuming an encounter between the Etruscans and the Romans, the crucial battle would usually be fought on flat land with mutually visible fronts that were not more than a kilometre or so long and often only a few hundred metres apart. Normally the two opposing phalanxes would simply head straight for each other, break into a trot for the last few metres, collide with a crash and then, drunk with terror and blinded by the dust, stab and shove till one side cracked. In this way, the issue was decided by a single, horrendous head-on collision in broad daylight on an open field, phalanx against phalanx.

The shock of the contending phalanxes would have been tremendous; several Greek authors familiar with this singular style of combat mention the crash when opposing phalanxes collided (e.g. Tyrtaios fr. 19.18, Aischylos Seven Against Thebes 100, 103, 106, Euripides Supplicants 699). The two opposing lines of death-dealing spears crossed (and often snapped), and the leading ranks were immediately thrust upon each other's weapons by the irresistible pressure from behind. It can be understood why the majority of the front rank of each phalanx went down in the initial crunch, 'While knees sink low in gory dust / And spears are shivered at first thrust' (Aischylos Agamemnon 66–7 Vellacott). However, their comrades stepped forward – or where pushed from behind – over their dead and dying bodies to continue the struggle, 'setting shield against shield they shoved, fought, killed, and were killed' (Xenophon Hellenika 4.3.19).

A word should be spared to consider the personal aspect of battle. The mêlée itself was a horrific, toe-to-toe affair, the front two ranks of opposing phalanxes attempting to jab their spears, with tips kept sharp by constant whetting, over their shields into the exposed parts of the enemy, namely the throat or groin, which lacked protection. What is more, these parts of the human body are areas of soft tissue in which serious and ultimately fatal damage can be inflicted. Meanwhile, with little room to do much else, the ranks behind would push. As can be imagined, once a man was down, whether wounded or not, he was unlikely ever to get up again. This brief but brutal mêlée was resolved once one side had practically collapsed. There was no pursuit by the victors, and those of the vanquished who were able fled the red field of slaughter. This was not, as before, a warfare with a dozen outstanding warriors, of idolized heroes. It was a modest and deadly warfare, the warfare of thousands of soldiers whose names we will never know.

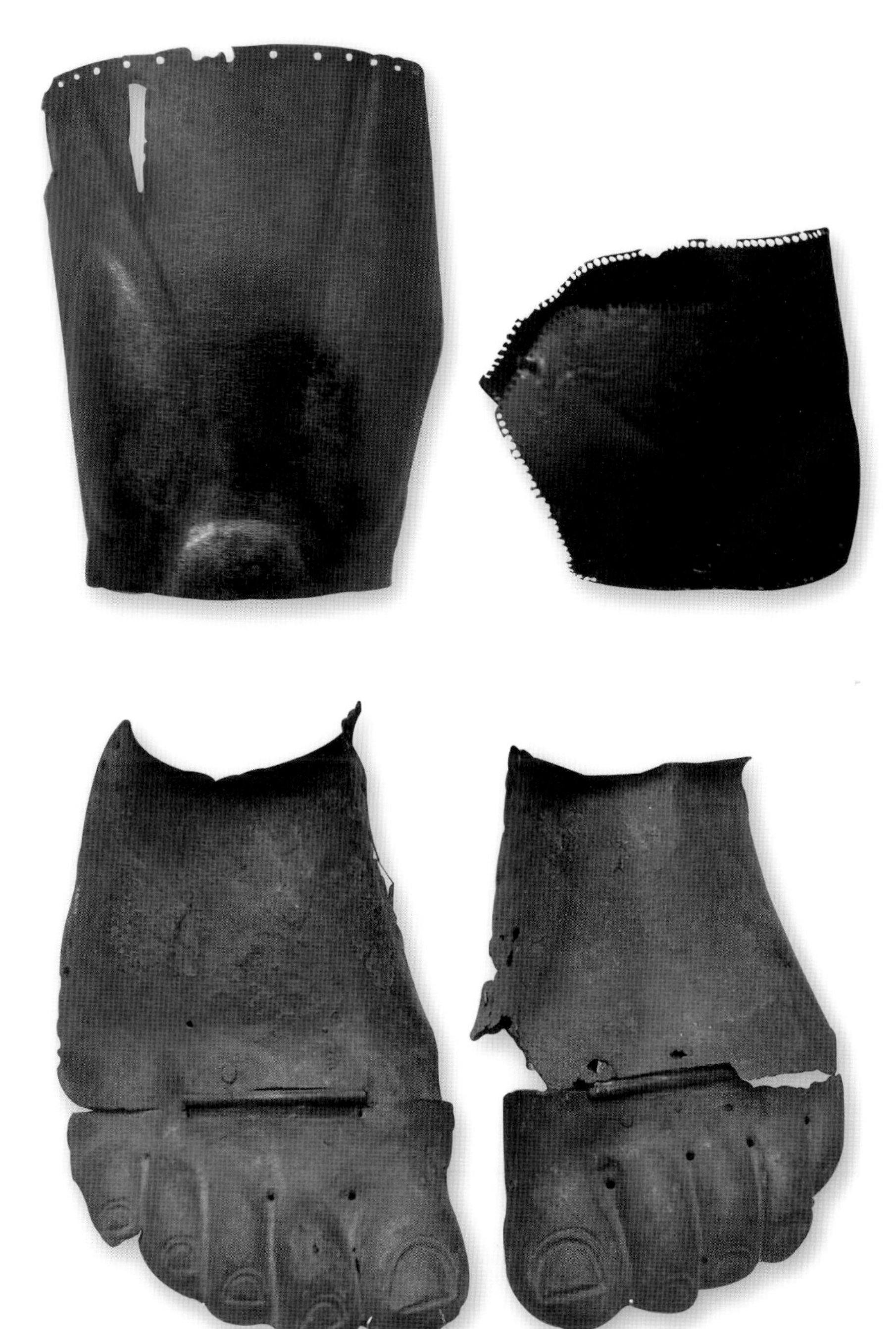

FAR LEFT
Bronze arm, thigh, ankle and foot guards, which were used in Greece in the 6th century BC, remained popular in Etruria for a much longer period. The thigh guard – probably the most useful extra piece of armour, especially before the common use of *pteruges* – was just an extension of the greaves and protected only the lower thigh. This example (Olympia, Museum of Archaeology) has the top part of the knee modelled onto it. It covered the front lower half of the thigh and was tied on with lacing. (Fields-Carré Collection)

LEFT
An ankle guard was a relatively simple piece of bronze moulded to cover the area of the heel, with the bronze coming over the top of the foot to be fastened by laces. There were embossed circles on each side to allow for the ankle bones. Perforations, clearly seen on this example (Olympia, Museum of Archaeology), allowed for backing to be attached. To have a guard for the ankle does seem odd, especially as it was a part of the body unlikely to be hit in battle. It has been suggested, however, that the ankle guard had much to do with the myth of Achilles' heel. (Fields-Carré Collection)

Pair of bronze hinged foot guards (London, British Museum, inv. GR 1856.12-26.714) from Ruvo, Apulia (*c*.520 BC). The extra pieces of armour fell out of fashion by around 525 BC, just as the *linothôrax* was being introduced. The one exception to this was the foot guard, which was in fact introduced at this time and, being awkward to wear, probably had more to do with parade than combat. Foot guards appear to have been popular with the Italian Greeks. (Fields-Carré Collection)

Polybios (12.25g.1) once wrote that it was nigh on impossible for a man lacking in the experiences of warlike pursuits to write about war. The same could certainly be said of the tactical component of warfare, the battle. It is within the arena of the battlefield that the nameless soldier witnesses the greatest violence in war. For him it is a wildly unstable physical and emotional environment: a world of boredom and bewilderment, of triumph and terror, of anger and angst, of courage and cowardice. It is a chaotic world that most of us are fortunately unfamiliar with.

DEFEAT, THE ALLIA, 390 BC

Italy in the Late Iron Age was a melting pot of different ethnic and tribal groupings. When Rome was an immature republic the brew was violently stirred up by the arrival in the peninsula of migrating Celts from western Central Europe (termed Gauls by the Romans), where the La Tène chiefdoms had emerged by roughly 500 BC. In the period from around 400 BC or thereabouts, land-hungry Celtic tribes (Boii, Insubres and Senones, amongst others) spilled over the Alps and colonized the Po Valley, evicting the Etruscans as they did so. From there they carried out forays against the heart of the Italian peninsula, far to the south and many leagues from the river Po. Totally indifferent to their own comfort, these were highly mobile pillagers whose sole objective was to get as much as they could, anywhere they could, and then head for home with the fruits of their summer's plundering once the rains of autumn set in. Their raids across the Po were a recurrent irritation.

It was a band of such energetic Gauls, Senones to be exact, that in 390 BC crossed the Apennines and swept down the valley of the Tiber, tossing aside the Roman army sent out to protect Rome, which was looted and burnt. Legend has it that the Capitol (a citadel as well as a religious centre) held out, but this is probably patriotic humbug, and the Romans, unable to save their city, were still obliged to buy off the visitors with a humiliating ransom. Fortunately for the Romans, these Gauls were primarily out for plunder, not for land, and they departed laden with their loot as suddenly and swiftly as they had appeared. In general such trips south were mostly raids carried out for the plundering of portable goods (wealth was still measured in gold and cattle), or for the purpose of securing prisoners for their ransom or for sale as slaves. Nonetheless, Rome was left poor and weak, and the morale of its citizens utterly shattered. It would take nearly half a century to recover from this Gaulish visit.

It was on the banks of the Allia, a stream that flowed into the Tiber just 11 Roman miles (16.25km) north of Rome and hardly a kilometre from Crustumerium, that the Senones utterly crushed the army sent to repel them. Livy states that 'the air was loud with the dreadful din of their fierce war songs and discordant shouts of a people whose very life is wild adventure' (5.37.8). Their war chief was one Brennus. He, very astutely, attacked the Roman right-wing reserves first, and when these broke before his charge the whole Roman army was seized with panic and took to its heels in a race for its life. The Senones, still hot on its heels, struck the Roman flank like a thunderbolt and drove the routers back to the Tiber. Here some lucky ones managed to escape across the river, but large numbers were cut to pieces. The disaster was shocking, the day, *dies Alliensis*, being forever remembered by the Romans as an *infaustus dies*, an unlucky day (Livy 6.1.11, Virgil *Aeneid* 7.717, Tacitus *Historiae* 2.91.1, Plutarch *Camillus* 19.1), though for us moderns it is a battle long elbowed into limbo by its more spectacular sequel, the sack of Rome.

GLOSSARY

Cuir bouilli — 'boiled leather' – leather soaked in cold water, moulded into shape, and dried hard using a low heat.

Hallstatt — Early Iron Age culture named after site at Hallstättersee in the Salzkammergut region of Austria.

Naue Type II — robust cut-and-thrust sword, originating from central Europe, whose flanged hilt had a distinctive curved outline where the rivets fastened the grip to the tang.

Pozzo/pozza — 'hole' – cremation burial consisting of a hut urn placed inside a larger clay jar, a *dolium*, which in turn is placed inside a stone-lined pit.

Situla/situlae — pictorially decorated bucket of sheet bronze.

La Tène — Iron Age culture named after site at La Tène on Lac de Neuchâtel in Switzerland.

Villanovan — Early Iron Age culture named after a necropolis site discovered (1853) at Villanova di Castenaso, a hamlet near Bologna in Italy.

BIBLIOGRAPHY

Ardent du Picq, C., 1903 (trans. Col. J. Greely & Maj. R. Cotton 1920, repr. 1946), *Battle Studies: Ancient and Modern*, Harrisburg: US Army War College

Barfield, L. 1971, *Northern Italy Before Rome*, London: Thames & Hudson

Carey, B. T., Allfree, J. B. & Cairns, J., 2005, *Warfare in the Ancient World*, Barnsley: Pen & Sword

Carmen, J. & Harding, A. F. (eds.), 1999, *Ancient Warfare*, Stroud: Sutton

Connolly, P., 1981 (repr. 1988, 1998), *Greece and Rome at War*, Mechanicsburg, PA: Stackpole

Cornell, T. J., 1995, *The Beginnings of Rome: Italy and Rome from the Bronze Age to the Punic Wars (c.1000 – 264 BC)*, London: Routledge

Cowan, R., 2009, *Roman Conquests: Italy*, Barnsley: Pen & Sword

Dawson, D., 1996, *The Origins of Western Warfare: Militarism and Morality in the Ancient World*, Boulder, CO: Westview Press

Dawson, D., 2001, *The First Armies*, London: Cassell

Errington, R. M., 1971, *The Dawn of Empire: Rome's Rise to World Power*, London: Hamilton

Fields, N., 2010, *Roman Battle Tactics, 390 – 110 BC*, Oxford: Osprey

Finley, M. I., 1977 (repr. 2002), *The World of Odysseus*, New York: New York Review Books

Forsythe, G., 2005, *A Critical History of Early Rome: From Prehistory to the First Punic War*, Berkeley/Los Angles: University of California Press

Forsythe, G., 2007, 'The army and centuriate organisation in early Rome', in P. Erdkamp (ed.), *A Companion to the Roman Army*, Oxford: Blackwell, pp. 24–41

Goring, E. (ed.), 2004, *Treasures from Tuscany – The Etruscan Legacy,* Edinburgh: National Museums of Scotland
Harris, W. V., 1979 (repr. 1985, 1986), *War and Imperialism in Republican Rome: 327 – 70* BC, Oxford: Clarendon Press
Keppie, L. J. F, 1998, *The Making of the Roman Army*, London: Routledge
Miles, G. B., 1995 (repr. 1997), *Livy: Reconstructing Early Rome,* Ithaca: Cornell University Press
Nillson, M. P., 1929, 'The introduction of hoplite tactics at Rome', *Journal for Roman Studies* 19, pp. 1–11
Oakley, S. P., 1985, 'Single combat in the Roman Republic', *Classical Quarterly* 35, pp. 392–410
Oakley, S. P., 1993, 'The Roman conquest of Italy', in Rich, J. and Shipley, G. (eds.) *War and Society in the Roman World*, London: Routledge, pp. 9–37
Paine, R. R. *et al*, 2007, 'Cranial trauma in Iron Age Samnite agriculturists, Alfedena, Italy: implications for biocultural and economic stress', *American Journal of Physical Anthropology* 132.1: pp. 48–58
Rawlings, L., 1999 (repr. 2009), 'Condottieri and clansmen: early Italian raiding, warfare and the state', in Hopkins, K. (ed.), *Organized Crime in Antiquity*, Cardiff: Classical Press of Wales, pp. 97–127
Rawson, E. D., 1971, 'The literary sources for the pre-Marian Roman army' in *Papers for the British School at Rome* 39, pp. 13–31 and *Roman Culture and Society,* Oxford 1991: pp. 34–57
Rich, J. W., 2007, 'Warfare and the army in early Rome' in P. Erdkamp (ed.), *A Companion to the Roman Army,* Oxford: Blackwell, pp. 7–23
Ross Holloway, R., 1994 (repr. 1996), *The Archaeology of Early Rome and Latium,* London: Routledge
Salmon, E. T., 1967 (repr. 2010), *Samnium and the Samnites*, Cambridge: Cambridge University Press
Sekunda, N. V. & Northwood, S., 1995 (repr. 1997, 1999, 2001), *Early Roman Armies, 600 – 300* BC, Oxford: Osprey
Smith, C. J., 1995 (repr. 1996), *Early Rome and Latium: Economy and Society* c. *1000 – 500* BC, Oxford: Oxford University Press
Smith, C. J., 2006, *The Roman Clan: The* Gens *from Ancient Ideology to Modern Anthropology,* Cambridge: Cambridge University Press
Stary, F. P., 1979, 'Foreign elements in Etruscan arms and armour: 8th to 3rd centuries BC' in *Proceedings of the Prehistoric Society* 45, pp. 179–206
Summer, G. V., 1970, 'The legion and the centuriate organisation' in *Journal of Roman Studies* 60, pp. 67–78
Warry, J., 1980, *Warfare in the Classical World,* London: Salamander

INDEX

References to illustrations are shown in **bold**. Plates are shown with page in **bold** and caption in brackets, e.g. **19** (18).